Cisterns: Sustainable Development, Architecture and Energy

RIVER PUBLISHERS SERIES IN EARTH AND ENVIRONMENTAL SCIENCES

Volume 1

Series Editors

MEDANI PRASAD BHANDARI
Akamai University, Hawaii, USA

and

Atlantic State Legal Foundation,
Syracuse, New York, USA

JOHN R. MATHIASON
Cornell University, USA

and

Associates for International
Management Services, Syracuse,
New York, USA

The "River Publishers Series in Earth and Environmental Sciences" is a series of comprehensive academic and professional books which focus on Earth and Environmental and Geo-Sciences. The series focuses on topics ranging from theory to policy and technology to applications.

Books published in the series include research monographs, edited volumes, handbooks and textbooks. The books provide professionals, researchers, educators, and advanced students in the field with an invaluable insight into the latest research and developments.

Topics covered in the series include, but are by no means restricted to the following:

- Sustainable Development
- Climate Change Mitigation
- Protected Area Management
- Institutional Architectures of Biodiversity Conservation
- Environmental System Monitoring and Analysis
- Migration/Immigration
- Flood Management
- Conflict Management
- Sustainability: Greening the World Economy

For a list of other books in this series, visit www.riverpublishers.com

Cisterns: Sustainable Development, Architecture and Energy

Alireza Dehghani-sanij

Memorial University of Newfoundland, Canada

Ali Sayigh

Director General of World Renewable
Energy Network (WREN), UK

River Publishers

Routledge
Taylor & Francis Group

LONDON AND NEW YORK

Published 2016 by River Publishers
River Publishers
Alsbjergvej 10, 9260 Gistrup, Denmark
www.riverpublishers.com

Distributed exclusively by Routledge
4 Park Square, Milton Park, Abingdon, Oxon OX14 4RN
605 Third Avenue, New York, NY 10017, USA

First issued in paperback 2023

Cisterns: Sustainable Development, Architecture and Energy / by Dehghani-sanij, Alireza.

Routledge is an imprint of the Taylor & Francis Group, an informa business

Publisher's Note
The publisher has gone to great lengths to ensure the quality of this reprint but points out that some imperfections in the original copies may be apparent.

While every effort is made to provide dependable information, the publisher, authors, and editors cannot be held responsible for any errors or omissions.

ISBN 13: 978-87-7022-939-5 (pbk)
ISBN 13: 978-87-93379-52-7 (hbk)
ISBN 13: 978-1-003-33757-7 (ebk)

Contents

Foreword

One of the important tasks which should be done at this period of the history in Iran is the revival of self-reliance and self-belief in people in general and in engineers and engineering students in particular. The social and economical progress in this country is possible if we believe that by relying on God and ourselves and cooperating with each other, we are able to eradicate all kinds of difficulties and barriers in pushing our way ahead. One way to develop our sense of self-reliance is becoming familiar with the methods employed by our ancestors to solve their problems.

Centuries before the discovery of electricity and the invention of air-conditioning systems of today, relying on themselves, our ancestors, had used their knowledge and abilities to provide comfort in their lives by creating Baudgeers (wind tower or wind catchers), domed roof, and yards; and to have cold drinking water in summer, by creating natural ice in winter and saving it for summer or saving cold winter water for summer. The scientific investigation of all these methods and designs proves the fact that our ancestors could have been able to create different types of masterpieces and we, engineers of this country, should be proud of what our ancestors have done in the realm of engineering and try to know these designs as much as possible and introduce them to others as well.

The book, *Cisterns: Sustainable Development, Architecture and Energy,* authored by my dear friend, Alireza Dehghani-sanij, is an attempt to introduce one of the engineering masterpieces of our ancestors to the world. He has meticulously investigated the thermal performance of Aub-anbars through the use of experimental, analytical, and numerical procedures. The last two Chapters of the book are devoted to the analysis of energy, exergy, and heat transfer in Aub-anbars, using artificial neural network models which are new methods in the study of thermal systems.

Alireza Dehghani-sanij is the author of several books all related to the Iranian masterpieces. Reading his books particularly, the present book is highly recommended for engineers and engineering students.

Mehdi N. Bahadori
Professor (emeritus) of Mechanical Engineering
Sharif University of Technology, Iran
January 2016

Preface

Through history, water and its different uses have played an important role in the development of societies. The importance of water usage is obvious and assessable from the ancient civilizations up to the present. Those in the early civilizations who lived in the arid and semiarid areas had to know much about water and the ways to access it in order to stop being nomads migrating all the time, and maintain permanent, secure, and safe settlements.

Since the beginning of life, water has played an important role in the existence of human race. It is the main ingredient of life. That is why the human beings have always searched for the sources where water exists. The theory of evolution started with Mancreation was born and developed and lived with the water.

The plateau of Iran is one of the ancient habitats of human beings. The earliest sign of man's existence on this land goes back to the tools of the Stone Age which are about 100,000 years old[1]. Different works from 4th and 5th millennia BC in different parts of this country have been obtained, which are the evidence of urbanism and the formation of minor cultures over 7,000 years old. In addition, some sections of the Iranian mythological history that is interwoven with its history events talk about 5,000-year-old Iranian civilization.

The man belonging to the plateau of Iran knew more than any other man how to confront the drought. His innovative ideas would have helped him to provide the needed water under any circumstances.

The need for water has pushed human beings to create engineering masterpieces, and Aub-anbar is an example of Iranian masterpiece. This structure, for more than thousands of years, with storing and cooling water has made it possible for the people to continue life in the hot seasons of the year. The architects and constructors of this engineering structure have been well aware of the principles of hydrology, hydraulic, thermodynamics, heat

---

[1]World Atlas of Iranian provinces, the research unit and Atlas authorship, under the supervision of Saeed Bakhtiary, Tehran: Geography and Cartography Institute, 2nd Edition, summer 2006.

transfer, fluid mechanics, statics, and mechanical of material and have applied all these principles in their works.

The book includes 11 chapters, and the title of each chapter is given in the table of content. In Chapter 1, the book discusses the indigenous knowledge and its importance for a sustained development and the key role the water plays in this regard. Chapter 2 talks about the sacredness and the position of the water in the culture, religion, and the civilization of Iran, and the structures related to Aub-anbar such as Qanats (or subterranean canals), dams, icehouses (or traditional ice-makers), and Sangabs (or stone troughs). Chapter 3 refers to the architecture of Aub-anbars. The hygiene and the quality of water in Aub-anbars are discussed in Chapter 4. Energy and its varieties, the importance of it in the environment, the development of societies and economy, and different ways of storing energy are discussed in Chapter 5. Chapters 6–9 are related to the experimental, analytical, numerical, and artificial neural network for heat transfer in Aub-anbars accordingly. Chapter 10 discusses the analysis of energy and exergy in Aub-anbars, and finally, in Chapter 11, some pictures of Aub-anbars in different parts of Iran are presented.

In the book, the creativity of the past can guide us to different aspects of existence in showing us how the old concepts can be used to solve some problems of our present life such as the water crisis, energy crisis, and environment protection. Aub-anbars can be used for providing cool drinking water and protecting foods. They can be of great use in earthquake, flood, bursting of water tubes, and drought as well.

To keep this native structure alive in the minds of people, other different purposes and usages are suggested here:

1. Cultural centers like libraries, amphi-theater, and fair grounds;
2. Museums (handicrafts, etc.);
3. Gymnasium (ancient body-building);
4. Traditional centers like coffee shops and tea shops;
5. Storage place for the water of fire departments in old neighborhoods;
6. Passive cooling of the buildings through canals;
7. Tourist visiting places.

The author likes to thank all the following who were of great help in one way or another to make this book materialized. Dr. Parviz Kardovani, Dr. Mohammd Hossein Papoli Yazdi, Dr. Ali Asghar Semsar Yazdi, Dr. Ali Arefmanesh, Dr. Ali Akbar Dehghan, Dr. Javad Safi Nejad, Dr. Farhad Fakhar Tehrani, Dr. Majid Rasekhi, Dr. Ahmad Sebt Hosseini, Dr. Mohammad Reza Khani, Dr. Kamyar Yaghmaeian, Dr. Ali Zafar Zadeh, Dr. Rajab Ali Labaf Khaniki,

Dr. Mostafa Khademi, Dr. Vali Anjil-ela, Dr. Mohammad Salmani, Seyed Masood Razavi, Mohammad Reza Haeri, Rouhollah Bagheri, Hossein Massarat, Mohammad Ali Mokhlesi, and ladies: Shaghayegh Sarafraz, Helia Sadat Hosseini, Nahid Safara, and Shabnam Navardi who studied the first draft of the book and made valuable comments and also I would like to appreciate Mohammad Hossein Dehghan, Hamid Reza and Mohammad Reza Dehghani Sanij, Vahid Pourgharib Shahi, Reza Ameri Siahoui, Saeed Vesali Barazandeh, Mohammad Reza Dehghani-sanij, Mohammad Reza and Mahdi Ameri Sefideh, Gholam Abbas Asadpour, Abd al-Aziz Eskandar Nejad, Hadi Karabi, Alireza Taheri Fard, Mahdi Nikbin, Ali Jalali, Hassan Naghashi, Masoud Godarzi, Shahram Zahmatkesh, Mohammad Ali Varang, Amin Dorost and Hadi Khazaee and ladies: Taha Norouz Zadeh, Mahin Mohammadnia, Zahra Beinaghi, Maryam Ramzali, Bahareh Parvaneh, Yas Naraghi, Sara Shamdani, Fatemeh Safaee, Naeimeh Rezazadeh, Niki Emami, Maryam Kasiri, and Tahreh Ghandali for sending pictures of different Aub-anbars and the managers and employees of central library under the auspicious of Cultural Heritage Organization particularly Setareh Eshaghi. My special thanks go to Dr. Khosrow Mehdipour, Dr. Fraidoon Tabrizi, and Sahar Tajaliwho kindly helped me with the final edit of the book in terms of language. Also, I would like to thank my great teacher Dr. Mehdi N. Bahadori for his insightful comments and his introduction.

List of Figures

List of Tables

Introduction

This book was written by Dr. Alireza Dehghani-sanij, Graduate Research Assistant in the Faculty of Engineering and Applied Science at the Memorial University of Newfoundland in Canada. The text was thereafter edited by Professor Ali Sayigh, Director General of World Renewable Energy Network (WREN) in the UK. The book is a follow-up to "Wind Towers" which was published by Springer in 2014.

Cisterns: Sustainable Development, Architecture and Energy was written strongly on believes that based on historical evidence and actual findings, Iran is most probably the country where cisterns, or *"Aub-anbars"* in Farsi, were first developed and built throughout the centuries. Therefore, it is quite natural for the author to name cisterns in the text *Aub-anbars*, as it has been called for centuries in his country, the translation of the same name having been used in other countries too. Although in some books, journals and papers published out of Iran by foreign and Iranian scholars, the names *Cistern* or *Water Reservoir* have been used. The word *Aub-anbar* is a compound noun in Farsi; *Aub* means *water* and *Anbar* means *tank/reservoir*. Putting them together gives the noun *Aub-anbar* and it should be used as one word.

People of the region wanting reserved cool water whether in cities or in different locations across the harsh desert during their travel. Queen Zubeida, the wife of Khalifa Haroon Al-Rasheed in 750 AD, built one of these cisterns closer to the town of Hiyal in Saudi Arabia so that the Pilgrims' Caravans going to Mecca will have cool, fresh water.

This book consists of 11 chapters with full analysis, illustrations, and photographs. It makes interesting readings to those interested in vernacular architecture, traditional buildings, and creative thinking.

Ali Sayigh
Editor
UK, February 2016

1

1

An Introduction to Indigenous Knowledge and Water Storage Systems

Geographical and climatic conditions have always affected the living style of people. The specific structure of the villages and towns in the arid and hot regions of the desert is the result of the climate in those areas. Burning sun, chilly weather at night, little rain along with high evaporation levels, strong winds, and noticeable temperature variation between sunshine and shadow in addition to the water shortage are important features of desert areas.

The residents of these regions have invented cooling and passive cooling systems to live in harmony with nature and maintain life despite the difficult climatic conditions. By using the systems and developing them through time, they were able to have cool air and potable water available despite the difficult geographical situation [1–3]. As a natural element, water is of the utmost importance in terms of living environment and the sustainability of the ecosystems, so it is not replaceable by any other element.

Located in the northern hemisphere in Asia and situated in the western part of Iranian plateau, Iran today covers an area of 1,648,195 km^2 known as a Middle East country. It is located between 44°5′ and 63°18′ geographical longitude and between 25°3′ and 39°47′ latitude. Because of its peculiar geographical situation and many scattered high and low lands and some other affecting factors, it is considered a dry region in the world. The rainfall in this country is less than 1/3 of the average rainfall of the world [4, 5]. The maximum amount of rainfall occurs in the northern parts and coastal region of Caspian Sea which is around 2000 ml annually, while in the west and northwest, the maximum rainfall is 500 ml, and in some areas, around 50 ml (Figure 1.1) [6]. In Figures 1.2 and 1.3 (a, b), climate and temperature Atlases are shown, respectively.

Figure 1.2 shows the Climate Atlas of Iran including cold and very cold areas, moderate and rainy and semimoderate and rainy areas, semiarid, hot

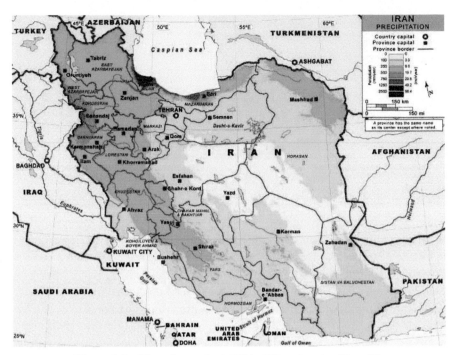

Figure 1.1   Map of annual rainfall in different parts of Iran [7].

and dry and very hot and dry areas, and very hot and humid areas. As it is illustrated in the above figure, a great part of Iran includes hot and dry, very hot and dry, and semiarid areas.

Figure 1.3 shows that in the southern areas, despite the humid air which exists throughout the region, the temperature level is high, so that in Khuzestan province, it rises to 54°C, and in Ahvaz (capital city), it reaches to 50°C [4]. In low central, eastern, and southeastern areas which have desert climate, it is extremely cold and extremely hot in winter and summer, respectively. The large difference of temperature level between night and day is another important factor that has to be mentioned.

In many places in Iran, water is saline containing lots of minerals and not drinkable. In some parts, there is not any surface or underground water available. That is why, in areas like the borders of deserts and dry plains where accessing surface and underground water is very difficult, great efforts have been made to provide and store water for many years. Developing

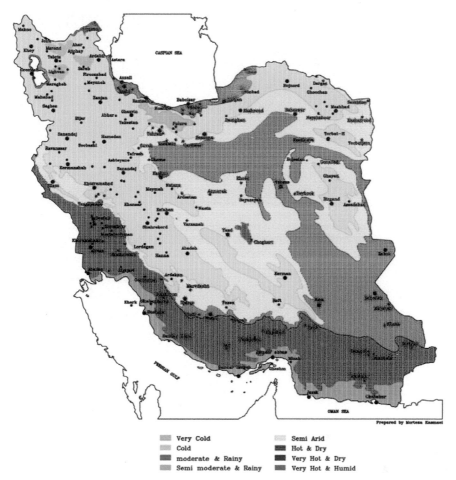

Figure 1.2 Climate Atlas of Iran [8].

Qanats (subterranean canals), dams, icehouses (traditional ice-makers), and **Aub-anbars** are examples of these attempts.

Aub-anbars (Cisterns) are indigenous systems for passive cooling and storing water. They have been used to store seasonal chilled drinking water in winter and partly in spring to be used in hot months of the year. Aub-anbars were entirely constructed based on the indigenous knowledge and technology.

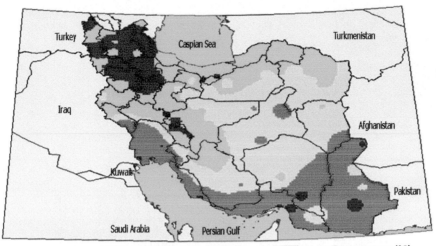

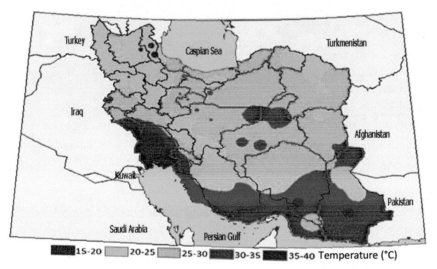

Figure 1.3   (a) Temperature Atlas of Iran in 2007: (1) average distribution of temperature in spring, (2) average distribution of temperature in summer [9].

1.1 Indigenous Knowledge

The key to success in achieving sustainable development and self-sufficiency lies in accessing water resources. In fact, real progress can be defined as balance between the goals of the development and those of the environment. It was

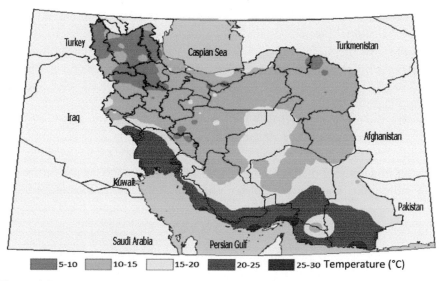

Figure 1.3 (b) Temperature Atlas of Iran in 2007: average distribution of temperature in fall [9].

not until 1980 that for the first time the expression "sustainable development" was mentioned in a report published by the World Organization for Protection of Natural Resources. This organization used the term "protection strategy of natural resources" to define the situation in which not only the development is harmless for nature, but it also helps in different ways. Sustainability may have four aspects: economic, political, social, and environmental. Actually, sustainable development does not focus on the environment; it also involves the social and economic elements as well. Sustainable development functions as the intersection of society, economics, and environment.

Indigenous methods being based on indigenous knowledge is known as the sum of the knowledge of the individuals which has been acquired through experiments and trial-and-error procedures. Based on that knowledge, technologies have been developed to fulfill the needs of the society. This knowledge has been passed from generation to generation through the years. Azkia states that "indigenous knowledge is the so-called knowledge that has been originated in a certain geographical region in a natural way" [10]. In another description, Taghvaee says that "indigenous knowledge is a cultural heritage and is the manifestation of the various aspects of the ingenuity of a nation." He further adds that "indigenous knowledge is the mysterious continuation of what a nation has created through the centuries and more things can be added to it in the future" [11]. Safee Rad also calls indigenous knowledge "the national wealth" including cultures, beliefs, values, methods, and tools. He also mentions that "this knowledge is accessible, cheap, and efficient. It has a holistic approach. It is called the cultural elements of a nation and its entire part... It gets stagnated in its birth place, does not develop and loses its applicability" [12].

Generally, the experiments driven from indigenous knowledge have been acquired within the framework of cultural, social, natural, and economic, situation of a certain region through centuries. Since this framework has always been changing and developing, flexibility and dynamism are in the nature of indigenous knowledge. On the other hand, the limitations of geographical conditions and inaccessibility of far reaching resources have made the indigenous knowledge innovative and cost effective as well.

The indigenous water-engineering structures, such as dams, icehouses (traditional ice-makers), and Aub-anbars are indigenous systems for storing water which have been formed according to indigenous knowledge having developed through the years.

1.2 The Importance of Indigenous Knowledge

Indigenous knowledge of each nation often makes it possible for the natives to fulfill their needs restoring to the natural resources without diminishing them. Therefore, the whole world's indigenous knowledge is a valuable treasure of the methods and tools experienced through the years which can be useful and efficient in the sustainable development of the nations in the world.

The applications of indigenous knowledge in a sustainable development can be summarized as follows [13]:

1. Safeguarding and preserving the natural resources through the indigenous managing methods, which is known as an appropriate model for the management of the natural resources in a sustainable development.
2. The success of the developing plans depends on the participation of the natives in all stages of planning, scheduling, implementing, and evaluating. Therefore, it is indispensable to use their knowledge.
3. Regarding the situation in the world, in many cases, it is necessary to mix the indigenous knowledge with the academic training to find the solutions to the problems. Due to the needs of world population, and the vulnerability of the remaining natural resources, neither of them proves effective by itself.

4. The goals of the development should be viewed through natives' eyes; the precise identification of the needs is only possible through effective relations with them and use of their knowledge.
5. In industrial developed countries, because of the domination of academic training in the process of production, the indigenous knowledge has almost been forgotten. To achieve a sustainable development in those countries, the application of appropriate indigenous knowledge from other countries and reviving the local know-how are essential.

1.3 The Necessity of the Full Attention to the Sustainability of the Water Resources

Since 1987, when Brantland Commission published its report titled "our common future," academic, scientific, and political associations have paid increasing attention to the concept of "sustainable development." Terms "sustainability," "sustainable use of resources," and "ecological balance" are widely used nowadays. In a sustainable development, water is considered the main axis, and the source of life. It is one of the three principal agents (air, water, and soil) in the formation and continuation of environment. Accessing water has been an important factor for the social and economic background and sustainability of culture and civilization. To achieve the sustainable development of water, it is essential to understand the cycle of water. Also it is important to know to what level the renewable sources of water should be consumed in order not to be diminished, and how it should be performed [14].

Today, most of the renewable water resources such as rivers, streams, relatively small lakes, and underground water which are quickly renewed and fed have been considered and taken into utilization. The cost of exploiting the water resources which are difficult to access is higher and the time needed to carry out the projects and to build the facilities to use the water is much longer. Besides, such projects cause heavy damage to the environment and man. Within the last decade, over 40,000,000 people had to migrate to other places because of the construction of big dams [15].

At the present time, using fossil fuels to pump water from deep aquifers of groundwater has brought us high volumes of drinking water at the cost of losing this capability in the future. The exploitation of underground water puts the durability of these recourses in danger. Too much consumption of underground water leads to many negative impacts on the environment, some of which are as follows:

1. Evacuation and reduction of the level of the groundwater. For example, in Ardakan, Yazd plain, the irregular consumption of the water has caused 0.5 m drop in the level of the aquifer of groundwater annually [16].
2. Land subsidence, an extensive environmental problem. Land subsidizes 0.5–1 m for each 10 m drop of water level. For example, in Shanghai, China, the level of the earth dropped 2.63 m during 1921–1965 which caused the water of Jiang Jiang River to get into the city and made a great damage. The entire infrastructure was ruined by the water [16].
3. Leaking of the seawater into the underground water in coastal areas deteriorating the quality of water.
4. Reduction of plant coverage.
5. The condensation of the alluvial texture and the reduction of its storage capacity.

As it is seen, the illogical expansion of water resources, within the last decades, has had negative environmental, social, and economic impacts. Most specialists have come to the understanding that the technology is not anything but the conversion of one kind of energy to another and the idea that any difficulty can be solved by new technologies is not always true. Too much exploitation of non-renewable energy resources such as oil, gas, and coal has caused environmental problems and air pollution. Therefore, diverse technologies should be designed and applied which can give durability and sustainability to the water resources in long run. Thus, avoiding the irregular exploitation of underground water, cleaning contaminated underground water, and avoiding the technologies which directly or indirectly can contaminate the water and any other activity that can interfere with the hydraulic cycle of an area are extremely important in viewing the sustainable development.

1.4 Characteristics of Sustainable Water Resources

The definition of sustainability includes four aspects: social, economic, political, and environmental aspects. Naturally, those systems would sustain and fulfill the objectives planned in all these four domains. Nevertheless, the sustainable environmental aspect has received more attention than the other elements. The extra attention to the environment has generated two principles:

1. A "sustainable system or a sustainable process" should be based on the resources that are not ruined or demolished in a relatively long run.
2. A "sustainable system or a sustainable process" should not cause an unacceptable level of contamination inside or outside the production unit [17].

Mir Abul Ghassemi states that the 1960s coincided with the water crisis in Iran. One reason for causing the crisis and intensifying it was disregarding the proper management and traditional methods of exploitation of water resources. He also adds that the success in long-term planning in a country depends on the features and abilities of the managers and the managerial methods which are driven from traditional culture of that country. The chance of success increases as the modern management complies and becomes more compatible with the skills and expertise of the local conditions [18]. He further mentions that 2/3 of Iran is covered with arid and semiarid lands. In those areas, the traditional management of the water resources has been handed from generation to generation and has evolved through time to the extent that within 3000 years, a great part of those deserts have turned into green gardens. He believes that "small-scale water plans is one of the characteristics of traditional management. Authorities should be aware of this fact because, at present, with the staggering cost of big water projects, the possibility of achieving the economic goals set for them is not much probable." Also, such projects have consequences regarding the sustainable development and preserving the living environment.

Mir Abul Ghassemi continues, "the residents of arid and semi arid areas have always been successful in finding innovative ways of using water resources and the reason for this success has been the implementation of small-scale projects for storage and use of water but in high numbers. Aubanbars were used in a great number to procure drinking water with the capacity of 300–3000 m^3 in deserts and dry widespread plains of Iran, and in cases up to 10,000 m^3 being excellent examples of those small-scale projects" [18]. He concludes that the use of limited resources and small water projects have been the secret of life in Iran and it can still be a key element for the success of water planning in this country. If in the previous projects, the primacy of the development and optimization of the traditional methods, particularly small-scale plans, had been preferred to the modern developing methods or used concurrently, the difficulties would have been less today.

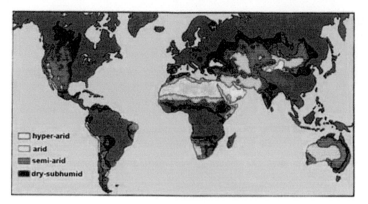

Figure 1.4   Arid zones of the world [20].

Considering what was mentioned before, the sustainable water resources are in close relation with the environment of the area in question, and they have a long-lasting age:

1. Can be implemented in small-scales and in widespread regions.
2. The local material can be used in these structures.
3. The construction can be done according to and compatible with the local conditions and expertise, no need for complicated technologies.
4. Most of all, the management can be based on the traditional culture.

1.5 Water Storage Systems in Different Parts of the World

Rainwater harvesting has been a popular technique in many parts of the world, especially in arid and semiarid regions (almost 30% of the earth's surface; Figure 1.4). Rainwater harvesting was invented independently in various parts of the world on different continents thousands of years ago. It was especially used and spread in semiarid areas where rainfall occurs only during some months and at different locations [19, 20].

The early storage systems were built in the Middle East. Researchers found some water storage structure which is estimated to go back to 9000 years ago in the mountainous part of Edom in Jordan [21].

In Yemen, ruins of dams and reservoirs as well as the unique, spectacular mountain terraces confirm the long history of water harvesting. The great historical Marib (Marieb) dam and its collapse are mentioned in the *Holy Koran*. Recent archaeological excavations (German Team, 1982–1984) discovered ruins of irrigation structures around Marib city dating back to the middle of the third millennium BC (some 4000 years ago) [22, 23].

Nevertheless, a great number of water structures belong to 1800–2500 BC which have been discovered in Palestine. In the loess plateau of China (Ganzu Province), rainwater wells and jars existed already two thousand years ago [20].

In North Africa, rainwater collection and storage are known to have been practiced during the 11th and 12th centuries. In Morocco alone, it was estimated in 1990 that there are over 360,000 Aub-anbars throughout the country that still supply domestic water to 10% of the population. In Egypt, some water-harvesting structures built in the Roman era have been cleaned and/or smoothed and put back into use [22].

As a representative of the Americas, I will say some words about the pre-Columbian practices of the Maya people in the Yucatan, Mexico. Mexico as a whole is rich in ancient and traditional rainwater-harvesting technologies (dating back to the Aztecs and Mayas). South of the city Oxkutzcab on the foot of the Puuc Mountain, we can still see the achievements of the Mayas [20]. In the 10th century AD, an integrated agriculture based on rainwater harvesting existed in this region. The people lived on the hillsides and their drinking water was provided by 20–45 thousand liter Aub-anbars called "Chultuns" (Figure 1.5 (a)). These Aub-anbars had a diameter of more or less 5 m and were excavated in the lime subsoil with a waterproof plaster. Above them, there was a ground catchment area of 100–200 m^2. In the valleys below (Figure 1.5 (b)), other types of rainwater catchment systems were used such as "Aguadas" (artificially dug rainwater reservoirs from 10 to 150 million liters) and "Aguaditas" (small artificial reservoirs containing 100–50,000 l) [20].

It is interesting to see that the Aguadas and Aguaditas were used to irrigate fruit trees and/or forests and to provide water for the plantation of vegetables and corn on small areas. Lots of water was stored, guaranteeing water supply even during unexpected droughts. This is one example of integrated water management, and we can find many similar ones all over the world [20].

In Iran, dam-like structures, "khoushabs"[1] (Figure 1.6), "houtaks",[2] and Aub-anbars, which are various systems for collecting water, date back to thousand years. 2000 years ago an integrated rainwater management and runoff agriculture existed in the Negev desert of Israel and Jordan [20]. In

---

[1]Khoushabs: Dams or open pits to collect rainwater and runoff water (in Baluchistan).

[2]Houtaks: Open water reservoir or tank to collect rainwater and runoff water (in Sistan).

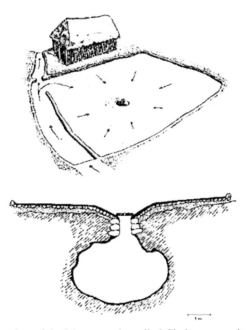

Figure 1.5   (a) Aub-anbar of the Maya people, called Chultun, capacity: 45,000 l, diameter: 5 m, catchment area: 150 m^2, the manhole is covered by a stone with a hole in the middle, where a wooden bolt is put, which recedes when it rains (Neugebauer) [20].

Figure 1.5   (b) Integrated water supply system of the Maya people in Xpotoit, Yucatan, Mexico (Neugebauer) [20].

Figure 1.6   A view of khoushab in Chabahar, Cistern and Baluchistan Province [24].

the Middle East, in countries such as Iraq, Saudi Arabia, and Iran, rainwater-collecting facilities have been observed on the route of caravans. These facilities would provide drinking water for the residents and the travelers.

On the Yucatan peninsula, the decrease in rainwater-harvesting methods was partially caused by the struggle between different indigenous peoples, but mostly because of the Spanish invasion in the 16th century. The Spanish colonizers introduced another agricultural system, various new domestic animals, plants, and European construction methods. These were not adapted to the environmental and cultural realities of the Yucatán. In India, similar reasons caused the disappearance of rainwater harvesting. The British colonial system was only interested in taxation, thus forcing people to abandon the village-based water-harvesting systems and causing the collapse of a century old system.

The technical progress of the 19th and 20th centuries occurred mostly in the so-called developed countries in moister moderate climate zones without a need for rainwater harvesting. As a consequence of colonization, agricultural practices from moderate climate zones were implanted in drier climate zones. Furthermore, emphasis was put on big dams, groundwater development, and piped irrigation projects with high input of fossil energy and electricity; this was another reason that rainwater-harvesting techniques have been set aside or totally forgotten [20].

Ferrokh and Sweden have investigated the traditional methods of water procuring in India. Traditional ways of water storage are regarded as the most important in India. The reason for the lasting of the old methods lies in the fact that those methods are compatible with the lifestyle in those places [25].

Indian villages are famous for their traditional water management. This includes, in particular, village tanks (also called village ponds), one of the most notable examples of riparian commons. There are between 1.2 and 1.5 million tanks still in use sustaining everyday life in the 660,000 villages in India. Tanks have been the most important source of irrigation in India [26]. Little rainfall, increase in population, and other factors have made people build those traditional water storages. At times when there is not enough water or there is drought, the reservoirs play a detrimental role in the economic and social life of the people. The positive role of the water reservoirs has been so immense that farmers before beginning to do agricultural activities in one area construct big tanks and reservoirs. This method itself has been a good reason for the growth and the development of farming around water reservoirs. Normally, it is not possible to dig wells in those areas.

Sri Lanka is a country where many villages have water shortage. Ariyanda and Secreatary state that "for the technical, managerial, and financial reasons, the storage of rainwater has been the only option feasible for the people in those villages" [27]. As far as the quality of the stored water is concerned, "it is better than the other areas in terms of physical and chemical features. And the filtered water has even better characteristics. To reduce the cost of structures, they are building in a way that is appropriate to the local features."

Andrew also believes that for achieving a sustainable development, water resources are of great importance. He states that "storing rainwater by using a simple technology and in a small scale is both feasible and cost effective and it greatly helps the preservation of nature and ecologic systems. This method is effective and sustainable for families especially in agriculture" [28]. Regarding the reports from Taiwan, to store the rainwater in that country, there have been nine million reservoirs built between 1980 and 2000 [19].

In the semiarid regions of Brazil, agriculture has been introduced in the recent decades. The local population had no possibility to experience rainwater-harvesting methods. As a result, they did not know how to live and work in a semiarid climate. But now, especially due to population growth and environmental degradation, people have to learn how to live in this rural semiarid region extending over 900,000 km^2. Reliable surface water is supplied for a very small area by the San Francisco River, and the groundwater in the mostly crystalline subsoil is scarce and saline. Therefore, rainwater is the most reliable source of water for humans and animals. From experiences of the past and other semiarid regions, we learn that the sustainability of water-harvesting systems is based on the combination of the basic needs of the farmers with the local natural conditions and the prevailing economic and political conditions of the region. People learn to live in a semiarid region by creating a new culture of relation to the environment and to water. This new relationship to the environment and water has been particularly fostered by the many grassroot organizations in the region. Nevertheless, there are still big irrigation projects underway along the Sao Francisco River. Large companies are planning to irrigate large areas to raise cash crops for export [20].

Alem has studied the water storing systems in Ethiopia and has concluded that the construction of rainwater-collecting systems has been an effort by people for survival against natural disasters and suggests the following [29]:

1. The rich indigenous knowledge in Ethiopia should be gathered and managed.
2. Government and non-government organizations should try to develop these systems in small scale.
3. The use of running water for home use and food production should be encouraged.

Garduro believes that rainwater-collecting systems are essential for the development of small villages in Mexico and states that "these systems are compatible with the ecologic-systems of arid and semi-arid areas and can reduce the unfavorable impacts of dryness" [30].

In Botswana, South Africa, the technology of rainwater collecting has been employed in schools and rural areas to collect water and store it in metal tanks. The technology proves that collecting surfaces are effective means for procuring the needed water of people. The efficiency of those surfaces has made them popular in rural areas [31].

Talebbeydokhti and Hooshyari have investigated the Aub-anbars in Fars province, Iran, and report that "Aub-anbars play an important role in the development and economic sustainability in an area. The water can be used for farming and prosper the agricultural business. Also the water of Aub-anbars costs less than the water of wells and dams" [32]. Aub-anbars are cheap and easy systems for procuring the water that can greatly decrease the rate of immigration. They also affect the social and individual behavior and strengthen the relationships by meeting around the Aub-anbar and discussing different issues. Thus, Aub-anbars have positive effects on social relations too. He concludes that "Aub-anbar is an appropriate system for sustainable development, and if with the modern technologies their shortcomings removed, they are the most effective means for storing water in some areas and their social and cultural impacts in previous decades have proved their effectiveness."

Javaheri brothers in one of the chapters of their book, *Water Solution in Fars,* have introduced the Aub-anbars of this region and call it a structure that has provided people with the water they needed through years [33].

Yavari in his book, *An Understanding of the Traditional Farming in Iran,* states that "Building dams and water streams, well-digging, and subterranean canals are the four main methods of procuring water in Iran. Meanwhile, water storage and irrigation by constructing Aub-anbars and reservoir would help water last for a longer time in dry seasons and consecutive droughts. In this way, the system is more flexible in the exploitation of scarce water" [34]. He refers to Aub-anbars as the complementary system of water resources and providing flexibility for the system.

Jafarpour and Motamed say that "collecting water and storing it in Aub-anbars and reservoirs, has been a way of life for men and animals as well. There are also so many diverse types of Aub-anbars which make the description of their details very difficult" [35].

Moradi has a view of water-collecting surface to Aub-anbars and states that "The methods to collect and use water in the ground are easier than the construction of big dams on the ground" [36]. Likewise, Ghoddousi states that "constructing big structures to procure water are impossible for several reasons and if possible they exert dangers to the environment like turning the downstream lands of the dam into deserts because of the allocation of the water to cities as drinking water and industrial units. To achieve a sustainable development, rainfall should be used where it falls. And Aub-anbar as the feature of a pond has this capacity" [37].

Nouzari has studied the biggest Aub-anbar in Iran, Kal Aub-anbar (Kal means a cut in the mountain in local language). He states that "Aub-anbars in South of Fars province have provided the crucial needs of that region. Creating Aub-anbars is the sign of Iranian ingenuity in the matching of man with the dry nature which is not kind with him" [38].

Rajaee Nia about Aub-anbars in Khorasan province says that "Aub-anbars or pools are the structures belonging to Islamic period which have been built like the other facilities on the main routes in hot and waterless areas. Aub-anbars with small area and deep into the ground were a safe spot for the travelers and farmers who had to trespass in those areas or did dry-farming. Aub-anbars were built near the seasonal rivers to be filled in rainy seasons and store the water for dry seasons" [39]. He also adds that "the design, material, distance, and the type of water resource, would be different according to the geographical condition; nevertheless, in most Aub-anbars some good principles of architecture have been observed."

References

[1] Dehghani, A. R. (2006). Study of history and evolution of wind towers, this beauty and nice phenomena and desirable traditional air conditioning system. *J. Air Cond. Refrig.* 19, 4–13.

[2] Bahadori, M. N. (1978). Passive cooling systems in Iranian architecture. *Sci. Am.* 283(2), 144–145.

[3] Bahadori, M. N., and Haghighat, F. (1988). Long-term storage of chilled water in cisterns in hot, arid reigns. *Build. Environ.* 23, 29–37.

[4] World Atlas of Iranian Provinces. (2006). *The research unit and Atlas authorship, under the supervision of Saeed Bakhtiary*, 2nd Edn. (Tehran: Geography and Cartography Institute Publications).

[5] Khani, M. R., Dehghani, A. R., et al. (2007). "Study of passive cooling systems and chilling systems rule in environmental pollution reduction," in *The 1st Conference and Exhibition of Environmental Engineering*, Tehran, Iran, 529.

[6] Khani, M. R., Yaghmaeian, K., and Dehghani, A. R. (2009). An experimental study in passive cooling systems and investigation of their role in diminishing energy usage and environmental pollutants. *Intl. J. Appl. Eng. Res.* 44, 519–528.

[7] http://www.atozmapsdata.com/zoomify.asp?name=Country/Modern/Z_Iran_Precip

[8] http://climaticdesign.net/?page_id=312

[9] Bahadori, M. N., and Dehghani-sanij, A. R., and Sayigh, A. (2014). *Wind Towers: Architecture, Climate and Sustainability*. Springer, Berlin.

[10] Azkia, M. (1996). Indigenous knowledge of temporary water exploitation in Yari Plain, Chabahar. *Jungle Grassland J.* 34, 4–10.

[11] Taghavee, H. (2000). The cultural-natural and tourism heritage: a survey in the interactive behavior, contacts and solutions in Iran. *Sofeh J.* 30, 26–43.

[12] Safee Rad, Z. (1998). "Indigenous knowledge and the cooperation methods in the preserving and reviving of natural resources, collection of articles," in the *Proceedings 1st Conference of Natural Resources and Cooperation in Development, People Office of Popularization and Cooperation*.

[13] Emadi, M. H., and Abbasi, E. (1997). Indigenous knowledge and sustainable development in villages. *J. Village Dev.* 1, 17–39, (2nd year).

[14] Ghodratmand, G. H., Bozorgzadeh, M., and Jahani, A. (1994) Water and population. *J. Water Dev.* 1 (Special Issue), 10–53 (2nd year).

[15] Asadi, M. (2000). The hygiene impacts of big dams. *Water Wastewater J. Isfahan*, 35, 55–61.

[16] Nanbakhsh, H. (1999). The impacts of too much exploitation of underground water on the environment and the measures to reduce them in Asian countries. *J Water Environ.* 35, 14–18.

[17] Speddings, S. C. (1996). *Agriculture and the Citizen.* London: Chapman and Hall.

[18] Mir Abul Ghassemi, H. (1996). Small plans, great need for sustainable development of water resources. *J. Water, Soil Mach.* 41, 3–7.

[19] Shirani, M. (2003). *Investigation of Sustainability of Indigenous Structure of Water (Aub-anbar) Storage in Lar Region, Fars,* MS Thesis, Management of Desert Region, Shiraz University.

[20] Gnadlinger, J. (2000). Rainwater harvesting for household and agricultural use in rural areas. *Presentation at the 2nd World Water Forum,* The Hague, Juazeiro, Brazil.

[21] Bruins, H. J., Evenari, M., and Nessler, U. (1986). Rainwater harvesting agriculture for food production in arid zones: the challenge of the African famine. *Appl. Geogra.* 6, 13–32.

[22] Naser, M. (1999). *Assessing Desertification and Water Harvesting in the Middle East and North Africa,* Policy Implication Discussion Paper No. 10, Center for Development Research (ZEF), Bonn, Germany, July.

[23] Abdulrahman, M., and Bamatraf, A. M. (1994). "Water harvesting and conservation systems in Yemen," in *The FAO proceedings of the expert consultation about water harvesting for improved agricultural production, water report3,* Rome.

[24] http://www3.bojnourd.irna.ir/fa/News/80471789/

[25] Ferrokh, L. and Sweden, C. (1998). *Indigenous Knowledge of Water Management,* 1998. Available at http://www.iirc.narod.ru/4conference/Section/sec4-1.pdf

[26] Pandey, D. N. (2000). "Sacred water and sanctified vegetation: tanks and trees in India," *Paper Presented at the Conference of the International Association for the Study of Common Property (IASCP), in the Panel Constituting the Riparian Commons,* Bloomington, Indiana, USA, 31 May–4 June.

[27] Ariyanda, T. and Secretary, H. *Rainwater Harvesting* as a *Water Supply Option in Rural Sri Lanka.* Available at: www.rainwaterharvesting.com/pdf/

[28] Andrew Lo, K. F (1999). *A Simulation Model of Flood Runoff Utilization in Taiwan*, Proceedings of the Ninth International Rainwater Catchments Systems Conference, Petrolina, Brazil, July.

[29] Alem, G. (1999). "Rainwater harvesting in ethiopia: an overview," in *25th WEDC Conference*, Addis Ababa, Ethiopia, pp. 387–390.

[30] Garduro, M. (2000). *Ancient and Contemporary Water Catchments Systems in Mexico*, 2000. Available at: www.cpatsa.embrapa.br/script/mnope.htm

[31] Gould, J. (1997). "Problems and possibilities relating to rainwater utilization in botswana," in *Proceedings of the 8th International Conference on Rainwater Catchments Systems*, Vol. 2. Tehran, Iran, 724–732.

[32] Talebbeydokhti, N., and Hooshyari, B. (1999). "Aub-anbars as a Trustable Method of Water Harvesting," in *Proceedings of the Ninth International Rainwater Catchments Systems Conference*, Petrolina, Brazil.

[33] Javaheri, P., and Javaheri, M. (1999). *A Solution for Water in Fars History*, Vol. 2. The Iranian National Water Treasure and National Committee of Irrigation and Drainage, Tehran.

[34] Yavari, A. R. (1980). *An Understanding of the Traditional Farming of Iran*. Bongah-e Tarjomeh va Nashr-e Ketab, Tehran.

[35] Jafarpour, E., and Motamed, A. (1991). *The Hot Environment of Desert*. The Research Center for Kavir and Desert Regions, Tehran.

[36] Moradi, N. (1997). Water Saving, Management of Water and Rain. *Salehin-e Rousta J.*, 136, Year 15, 2–14.

[37] Ghoddousi, J. (1998). Environment means water, means life. *Ravesh J.* 34, 4–6.

[38] Nouzari, M. (1992). Kal pond, the biggest Aub-anbar in Iran. *J. Geo. Res.* 7, 134–139.

[39] Rajaee Nia, M. A. (2002). Aub-anbar (Darfak), a Spring in Kavir. *Jam-e Jam Newspaper*, 3, 7.

2

The Indigenous Water-Engineering Structures in Iran

Water in the history of mankind has always been considered a sacred and civilizing element. Wherever development and prosperity has been shown throughout the world, it owes itself to this life-giving precious gem. There are many tales in the Iranian history of culture and civilization regarding the sacredness of water, with themes like respecting the water, not wasting it, and not contaminating it. The reason is that water has a deep root in the culture of Iranian people since the ancient era.

For a long time, the plateau of Iran has been faced with the lack of water, and the residents have tried to confront the water shortage in different ways. In the evolution of the civilization in the widespread land of Iran, this trouble caused many water-engineering structural masterpieces to be created which could give courage to people to stay and live on the land (Figure 2.1).

Water-engineering structures playing an important role in the daily life of people have gained a high position in the culture of Iranians, so much so that they became sacred. In order to collect and use the water in the arid and semiarid lands, the clever and dexterous Iranian engineers and architects made developing Qanats (or subterranean canals) its own profession. In the areas like Khuzestan and Fars provinces where permanent and seasonal rivers were running, they erected dams. They built Aub-anbars to store the rainfall and icehouses (traditional ice-makers) to produce ice. Sangabs (stone water container troughs) for social and religious reasons were usually placed in religious locations like mosques and the like for the trespassers to quench their thirst.

Figure 2.1   A small statue made of lime stone, showing a man holding a water jar. It was found in Shoush, Iran. It belongs to 2800 BC displays in Louvre Museum, France [1].

2.1 Sacredness and the Position of Water in the Culture, Civilization, and Religion of the Iranian

In ancient Iran, water was considered a sacred substance and people believed after fire, it was the most sacred element and the second creation out of seven which 'Ahura-Mazda' has created. Ancient Iranians considered the spring, river, snow, and rain sacred too and respected them. In Zoroastrian doctrine, safeguarding water from dirt has been one of the principles of this religion, so that "Eizad Nahid," "Ardavisur Anahita" (the strong clean river), was the Goddess who guarded the water and they considered her a reputable goddess in Iran (Figure 2.2). People made many temples to worship her [2–5]. According to traditional rituals, the 31st of September was called "Abangan Day," a day devoted to water. There were other Gods and Goddesses, such as "Aban," "Apam Nabat," and "Tishtar," who had a share in safeguarding the water on earth and they were worshiped as well.

In "Avesta," the holy book of Messenger Zoroaster, "Aban Yasht" (fifth chapter), is one of the ancient and enchanting Yasht which describes "Anahita." Some parts of this Yasht say:

"I worship the water which is the creation of Ahura-Mazda"

"Warm glory is for the pure life-giving water and warm glory for 'Ardavisur Anahita', the praiseworthy guardian of non-polluted waters on the Earth"

Figure 2.2   "Anahita," stone cut in Sassanid Period, Tagh-e Bostan.

The connection between water and rituals continued in Iran after Islam too, so that the worshiping places of "Nahid" turned into glorious praying places outside the cities where Iranians would gather together and pray to God to send them rain [3–5].

Islam also calls water the origin of life, and there are several verses in the *Holy Koran* about water [6]:

"...And from water we gave life to everything ... " (al-Anbiya: 30)

"...So we sent water from the sky ... " (al-Hijr: 22)

"We sent water from the sky, it ran, and the rivers had their share of water ... " (Raad: 17)

"And the God who send rain from the sky, enough water to give life to the dry and dead desert ... " (Zokhrof: 11)

2.2 Qanats (Subterranean Canals)

Any newcomer to the Plateau of Iran Oases of[1] Sahara in the North Africa would observe circular holes with average diameter of 1 m on the ground. They start from the foothills of mountains, and after passing through arid and semiarid deserts, they reach the green villages and towns where people live. Those holes, spread across the deserts of Iran, are in fact the well-like vertical shafts which connect the surface of the ground to the underground tunnels. Those tunnels carry the water from the high land to the residential areas at the low plains. Qanats are composed of a number of underground vertical wells called shafts and one or more approximately horizontal wells called corridors, with a slope less than the slope of the ground surface. The structure carries the underground water in the mountains and relatively high areas to the low plains by means of gravity force. The vertical wells are used for the digging of the underground tunnel, as well as repairing and maintaining (dredging) of Qanat. A sketch of a Qanat with the shaft and corridor is shown in Figure 2.3.

The age of the Qanat, which was invented by Iranian well diggers, goes back to thousands of years, and is rooted in the ancient history of Iran. According to Henri Goblot, Qanats were not a technology of irrigation in the first place; it was a system made to get rid of the troubling underground water while excavating the ground for mining [1, 7, 8]. One of the Iranian kings, Manuchehr from Pishdadian period, ordered that the skill of connecting wells to each other and the carrying of water through them should be taught to the farmers. From the early 1st century AD, digging wells by Maads and Sassanids developed across the plateau of Iran. Qanats have played an important role in the urbanization of, and procuring the water for, great cities such as Ekbatan, Pasargad, and Persepolis. Then, the change of rulers and the economical and social conditions brought ups and downs in the development of Qanats but it never stopped.

Quoting Antiochus III of Seleucid, Polybe (210 BC), and the Greek historian writes [1, 7, 8]: "In fact, no where is seen water on the surface of the ground. Water is running through the underground canals and feeds the reservoirs which are unidentified for the people who do not know this land. People of this country, work hard and spend a lot, lead the water from far distances to the areas they need. So that today, those who benefit from the water do not know where the origin of the water is." In the period following the arrival of Islam, Qanats were still playing the important role of procuring

---

[1]A small fertile or green area in the midst of desert [9].

water for the villages and towns. A role that is still the less in some areas of Iran be played.

The technology of building Qanats spread from Iran, as an important intersection for the ancient nations, to the other parts of the world. The technology was transferred to Oman, Masqat, and Saudi Arabia by Kourosh the Kabir around 525 BC. Around 500 BC, by military campaigns and the order of Dariush the great, Qanat was transferred to Egypt. After Islam, around 750 AD Qanat technology was gone around to the north of Africa and Spain. The Qanat of "Yafuga" was built by Muslims in Madrid, the capital of the Spain, at the same time. In 1520 AD, Spanish built Qanats in Mexico, and from there, it was transferred to Los Angeles in northern America. In 1540 AD, the city of Pica in Chile owned a Qanat. The spread of Qanat technology in East has an old history. Figure 2.4 shows the emergence and the spread of Qanat know-how throughout the world.

At present, besides Iran, 36 other countries have Qanats. The names given to Qanats in different countries are not the same. Totally, there are 27 names for Qanat: They are "Kariz" (Iran), "foggara" or "fughara" and "Khettara" and "Iffeli" (North Africa), "Kariz" or "Karez" (Afghanistan, Pakistan, Azerbaijan and Turkmenistan), "Falaj" or "Aflaj" (Oman), "Ain" (Saudi Arabia), "Kahriz" (Iraq), "Kanerjing" (China), "Foggara" (Algeria), "Khattara" or "Khettara" and "Rhettara" (Morocco), "Galleria" (Spain), "Qanat Romoni" (Syria and Jordan), "Galerias" (Canary Islands), "Mambo" (Japan), and "Inguttati" (Sicily). Some other terms used for Qanats are Ghundat, Kona, Kunut, Kanat, Khad, Koniat, Khriga, Fokkara, etc. A great number of Qanats in Iran are still functioning.

The deepest Qanat probably dated back to the arrival of the Aryans is the "Ghasabe" Qanat in Gonabad northeastern of Iran, which is 70 km long (Figures 2.5 and 2.6). Mother well of this Qanat is about 250 m deep.

The longest Qanat in Iran is "Zarch" Qanat in Yazd. The history of this Qanat dates back to over 1000 years, is around 80 km long, and has 2115 shafts. The most surprising Qanat in Iran is a double-deck Qanat called "Moon." It was built in Ardestan in the margin of central Kavir of Iran around 800 years ago. Among other interesting Qanats discovered in 1992 in Iran is Kish Island's Qanat which is 2500 years old. So far, 200 shafts have been discovered that are connected to this Qanat. The distance between shafts is 14–16 m. The ceiling of the Qanat is covered with coral layers, 2–15 m thick. There is a running Qanat called "Ghaza Ghurtaki" in Ghazvin, with a plant growing in its entrance with curing effect [7, 10]. The conclusion drawn from the foregoing discussion is that Qanat is one of the traditional technologies that can carry the water

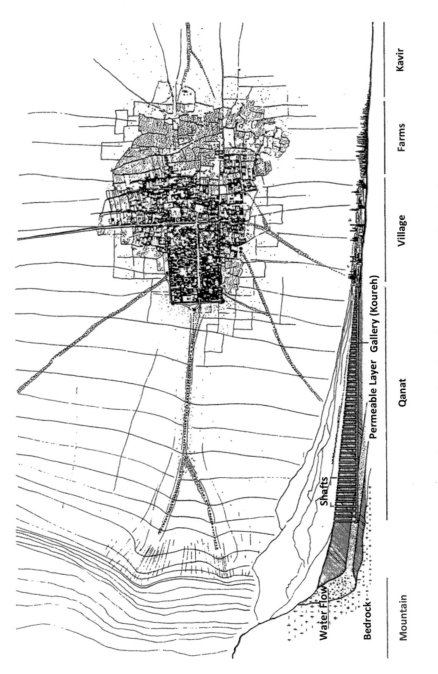

Figure 2.3 A view of the Qanat leading water to a village [1, 7].

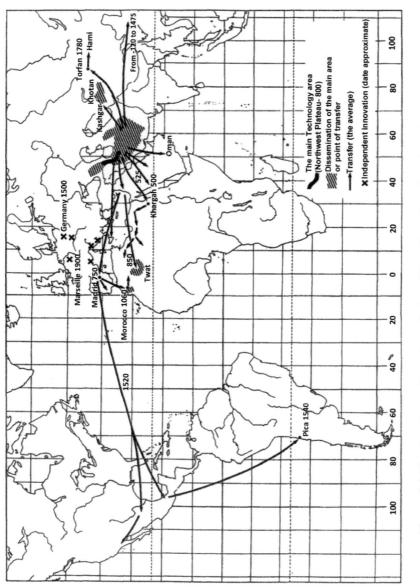
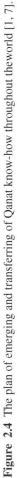

Figure 2.4 The plan of emerging and transferring of Qanat know-how throughout the world [1, 7].

Figure 2.5   A view of the horizontal corridor of Ghasabe Qanat, Gonabad, Khorasan Razavi Province [12].

Figure 2.6   A view of the origin of the Ghasabe Qanat, Gonabad, Khorasan Razavi Province [12].

from far distances to the villages and towns to provide them with drinking and irrigation water, while posing no threat to the environment, and more importantly, it does not need to consume any non-renewable type of energy.

One thousand years ago, an Iranian scholar called al-Karaji, in a thesis titled *Excavation of the Hidden Water*, discussed the ways of finding water, building Qanats, leveling, investigating the quality of the water, bounds of Qanats, and similar issues. Most of his ideas are still valid [11]. The indigenous Qanat engineering, including the method of digging and management, is a sign of the knowledge of ancient Iranians in the old days, a great knowledge that has been handed throughout centuries from one generation to another.

2.3 Dams

Dam construction is an engineering activity in which the historical and geographical conditions of the areas play a significant role in their formation and spread. In the past, due to the necessity and needs of the residents in any specific geographical area, a dam or pond would be built to supply the area with the needed water. In some other areas, due to the low water level of the rivers and streams or the need to divert the path of the river, a dam was built in order to increase the water level and use it for the agricultural purposes.

Any kind of obstacle in the path of the water which raises the water level and lets the water be stored is called a dam. In other words, a dam is a barrier which is placed across a stream or river to permit a higher water elevation and storage [13]. A dam could just be constructed for a specific purpose (mono-purpose dams) or a combination of several purposes (multi-purpose dams). The following could be the purpose of constructing a dam:

1. Procuring the drinking water;
2. Procuring the water needed for agricultural activities;
3. Controlling the floods and regulating the flow of streams, rivers, and flood water;
4. Water power plant;
5. For industries and their specific needs;
6. Diverting the flow path of the water by raising its level;
7. Shipping and transportation;
8. Preserving the environment and creating recreational centers.

In countries such as Iran and Egypt which have always been prone to the floods from the very old days, dams are constructed alongside the streams, rivers, or flooding areas to prevent the damages of the overflowing water. The history of dam building in Iran, Egypt, and Mesopotamia goes back to a thousand years ago, and the signs of old dams in these areas can still be found. Generally, dam building, dredging, and repairing were carried out and managed by governments and the kings who were enthusiastic about development and prosperity. The economic prosperity and progress of villages and towns greatly depended on the dams and the importance rules would attach to the dam-building and similar water-related structures. Dam building was very much developed in Ache median era. The geographical condition of the country on one hand, and the interest of the kings on the other, helped the development of dam building in Iran. Too many dams were built in the southwestern parts of the country during this period. Many water supply systems and irrigation facilities which had been used for many years owe their existence to the efforts of Iranian engineers and craftsmen. The remains of those structures are still observable in different parts of the country, as well as in the countries subjected to the ancient Iran. It was in the Ache median period that the first attempts were made to build a dam across, Arvand and Forat Rivers. One of the old rivers connected to Arvand was "Diyala" River, and Cyrus the Great ordered a dam to be made from soil and wood across this river which fed the irrigation network. Apart from the dams built on Arvand and Forat, there were several dams built across "Kor" River in Fars for the irrigation of the lands around Persepolis. Although there is not enough evidence of all the dams built in Ache median period, the remains of some dams on those rivers reveal that the columns of the dams belong to that period. An example of this is the remains of the "Naseri" dam located 48 km northwest of Persepolis. Ibn-e Balkhi (born in 850) describes this dam as [14] "In this part of the river, a dam was built in old days which would furnish the water needed for the lands around. It was ruined when Arabs attacked Iran, and there was no more farming."

In the Sassanid period, and the ruling of Shapour I, the emperor of Rome, valerian, was defeated by the Iranian army and taken into captivity. Shapour used the Roman soldiers to help build some structures in Iran. One of those structures was the "Shadravan" dam in Shoushtar across Karoon River. Shoushtar, which is located on the eastern side of Karoon on stone banks, was considered an important city in Sassanid period.

Another dam, which was built after Shadravan in Shoushter, was Ahvaz dam where the remains can still be seen. That dam was 1000 m long and probably 8 m wide. Still another dam which was built in 4 AD ordered by Shapour II or Ardeshir II is the bridge-like dam of Dezfoul that was built on Kofe River. There is the remains of another dam called "Ghir dam" on Karoon where the two rivers of Gorgor and Dez join together and was regarded the third important dam in the area on Karoon.

The Sassanid developed the watering system of Diyala River so much so that the water in the river was not enough to feed all the branching streams. They had to use the water of Arvand River. For this, they had to raise the level of the water in this river first by special raising tools and then leading the water to the Diyala River through canals. The water systems and its development in the south of Iran reached its peak in the time of Khosrow I, Sassanid King (531–579 AD). One of those developments is the Nahravan canal which was fed with the water collected behind the dam on Arvand, near a place called Dur. That canal was repaired during Abbasid rulers. The Nahravan canal in Badkobeh also (a place at a distance of 53 km from the northeast of Baghdad and about 110 km from the downstream of the dam) reached to Diyale River.

The interesting point is that the Nahravan canal and the Diyala River were at the same level and they could join each other without any virtual control. This shows that the Sassanid engineers could have chosen the location of the dam in a place where there was no need for any man-made structures to do that and they were skillful designers.

About 36 km away from the south of Badkobeh, a dam was built called Beladi to control the Diyala River. Water from this river flowed to a short canal and poured into Arvand, somewhere below Baghdad and above Tisfoun. In addition to the bridges and the dams which were explained, there were several other bridges and dams built in Khuzestan in the distant past, which were a great help for irrigating of the lands around. Some of these dams are called Ghale Rostam Dam, Shoeibieh Dam, Karoon Dam, Ajirab Dam, Karkheh Dam, Abu al-Abass Dam, Abu al-Fares Dam, Jarrahi Dam, and Eizad Khast Dam [13, 14].

2.4 Icehouses (or Traditional Ice-Makers)

In the past, icehouses or traditional ice-makers were other complementary structures to Aub-anbars used by people. The structures used passive cooling methods and took advantage of low temperature in the freezing cold nights of the winter, to produce ice. The founders and creators of these structures, by using their talent and simple architectural elements, developed a collection of icehouses in the margin lands of villages and towns (Figure 2.7) [1, 7].

There is not a precise history of icehouses available up to the Safavid period. Although in many instances before Safavid era, reference to ice had been made in historical texts and even in poems, there had not been any mention of how that ice was made. The term "ice" is abundant in the poems of many Iranian poets. Also, references have been made to "icehouse" [15] in historical texts. Based on the documents available, Jalali Icehouse is the oldest structure of this type from the old days. Figure 2.8 shows the picture of an icehouse which John Chardin, French tourist (born in 1643 AD), brought with him from Kashan.

Building technology and the architectural method in the structure of ice-makers are all signs of the precession and insightfulness of those builders and architects to important points such as isolation and preserving the appropriate coldness to keep ice, the continuous parting of water from ice, building materials, and how to provide ice [17]. The structure of the icehouses is not very complicated. Icehouses are made up of three parts [17–20]:

1. Shadowing walls
2. Ice-making pools
3. Ice-reservoir (the main ditch for ice).

The shadowing wall was a very long and tall wall which was extended from east to west. Those walls sometimes reached 15 m in length. They saved the frozen water from the sun shine, and if there was a little sun shining on the ice surfaces from the east and west, shadowing sidewalls were also built which were connected to the main wall (Figure 2.9).

The ice-making pool was a rectangular ditch which was dug parallel to and at the side of the shadowing wall outside the building at the northern part of the structure. Its length was less than the wall's length, and its depth was 30–60 cm. Those pools were the place where the ice was made in the cold winter nights, and in most cases, there was no material used in their construction.

The main ditch was the place for the storage of the ice and it was dug on a level ground. The walls of the ditch were usually covered with mud mixed with straws or coated with a special material (*Sarooj*), and there was a narrow canal connected to a deeper well. The melted water of the ice flew to that well away from the main ditch [23]. The buildings erected on the main ditch were dome-shaped or tunnel-like (Figures 2.10 and 2.11) [17].

To provide ice, the ice-making pools were filled to a certain point in the cold winter nights. The freezing weather would freeze the water inside the pools. A certain volume of water was led over the ice mass in the following nights. This amount would freeze during one night. Generally, the height of the water would not exceed more than a few centimeters and this was repeated several nights until the whole ditch was filled with the ice. Then, in the morning, people broke the ice into pieces and carried each piece, using chains, to the ice-storing place and stacked the pieces upon each other.

When it got hot and the necessity of using ice was felt, some people wearing boots and using iron hooks, pulled the ice bars out of the ditch and took it to a large scale, whose pans resembled a pair of door and was hung from the roof by four long chains in the adjacent room. Those rooms were only opened in the morning or in the evening when the sun was setting, and the ice was broken into pieces and taken to the market to be sold.

Figure 2.7   A view of Moayedi Icehouse in Kerman Province, Iran.

Figure 2.8   A view of Kashan and its icehouse beside the surrounding wall of the city [16].

Bricks and mud were the most important materials used in the desert structures, particularly in the construction of the icehouses, because they are not only easy to be procured, but they are also the best isolation for preventing the penetration of heat from outside to the internal space. Raw mud–bricks and the mud covering mixed with straws are the most resistant materials in Kavir. A plaster made of clay and straw is an excellent insulator against rain and snow water. The khaki color of the material reduces the intense reflection of the sun rays and gives a pleasant view to the structure.

Stone and bricks are among the important materials because of their availability and inexpensive price, which were used in the construction of icehouses. Stone was usually used in the foundation and bricks in the ceiling. In such cases, the exterior of the icehouse was certainly covered with a mixture

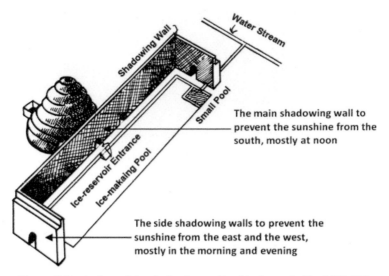

Figure 2.9 A view of the shadowing walls of icehouse in Yazd [21, 22].

Figure 2.10 A view of a dome-covering shape upon the main ditch of the icehouse in AbarKooh, Yazd Province.

of the clay and straw. Also, the walls of the icehouse were built with the stone and bricks and were covered with a mixture of clay and straws (Figure 2.12) [17, 19, 20].

Figure 2.11 An interior view of an icehouse with nine steps in Urmia, West Azarbaijan Province [21].

Figure 2.12 A view of an icehouse with its walls in Zavareh, Ardestan, Isfahan Province.

2.5 Sangabs (or Stone Troughs)

A few centuries ago, most of the time, water was carefully mixed with rose water, the aromatic essence of fruits or syrups, and the resultant water which was cool and fragrant was kept in large pots made of stone called Sangabs to be given to thirsty trespassers. For this reason, in 17th century AD in the

Figure 2.13   The Sangab at the entrance of Imam Mosque in Isfahan.

Safavid, Sangabs were made by scraping voluminous stones and decorating with regular, delicate geometric designs. Such designs would reduce the harshness of the stone and gave it a softer view (Figures 2.13 and 2.14). The margin of the Sangabs was adorned with wide, long poems, and calligraphy the concept of which was generally the praise and gratitude to the God who has bestowed life with water (Figure 2.15).

The present Sangabs are in different shapes. The commonest shape is hemispherical in two types. One type is secured on a support but the other one is not. Most of the Sangabs of Jam-e Mosque of Isfahan are of the type without support (Figure 2.16). The other shape of the Sangabs is rectangular. In some, the length of the stone is about 1.5 m and the width is 0.5 m like the one in the entrance of Prince Ibrahim in Isfahan.

For religious, social, and charity reasons, Sangabs were usually placed in mosques, shrines, and public locations such as passageways, the intersection of alleys, bazaars, and similar ones. There was a small metal bowl with beautiful decorations on it upon the Sangabs used it for drinking the cold water.

2.6 Aub-Anbars

As mentioned before, Aub-anbars are one of the indigenous systems for passive cooling and storing water. They were used to store cold drinking water in winter and part of spring. The stored water would be consumed in the hot seasons of

Figure 2.14   The Sangab in Chahar Bagh School in Isfahan.

Figure 2.15   The engraved writings of on Chahar Bagh Sangabs in Isfahan.

the year. Aub-anbars are in fact covered and isolated reservoirs which were built in lower levels than the ground surface or in the mountain areas. The main function of Aub-anbars was to provide chilled water for the towns, villages, fortresses, caravan routes, and caravanserais and to store water for the time there was no rain or droughts would occur and when there was a war [24, 25].

Figure 2.16   The special designs, engraved on the Imam Mosque Sangabs in Isfahan.

In Iran and some other countries, the concept of Aub-anbar has been expressed with various terms such as pond, reservoir, pool, and Barkh (Iran), Maśna (Egypt), Sardāba (Turkmenistan), Khazzan (Palestine), and Hauz (Afghanistan) [2, 4, 25].

The water reservoirs in the beginning were only ditches which would be filled with the rain water and flood. Gradually, people themselves tried to store water in them and finally made the reservoirs in appropriate locations. With the progress of social life, the methods of water storage became better and better. In countries like Iran, Egypt, and Mesopotamia, water was stored in covered isolated reservoirs where water would not be evaporated or contaminated [26]. In Greece and Rome, this same method was performed to store water [27]. Some of the Aub-anbars in Palestine date back to Nabatian Period [28].

Abbasi knows Aub-anbars as old as Qanats [29]. He states that Aub-anbars were a way of confronting the shortage of water in arid and semiarid areas. Moradi thinks of Aub-anbars as collecting surface water and says [30]: "Aub-anbars and Qanats were built in Kavir and semi-arid areas of Iran from the very distant past and Iranians have been among the earliest to innovate such structures." Ghobadian also states that [22] "In Iran, Aub-anbars have been a means of storing water in old days and probably the Iranian may be known to be the creators of Aub-anbars in the world."

The record of Aub-anbars construction in Iran, despite its different design, certainly goes back to several thousand years ago. The oldest Aub-anbar discovered is the reservoir of Dur Untash in Chogha Zanbil, Khuzestan Province which dates back to 1250 BC [25]. From Archimedean period, there are several Aub-anbars seen in Persepolis [22]. One of the oldest Aub-anbars still existing is Persepolis (Takht-e Jamshid) well which was dug in the period of Darius I [31]. This well had been dug in the heart of hard and compressed stones in the skirt of a mountain. There are other ponds left from Sassanid era in the south of Iran and the islands of Persian Gulf [2]. There are some old ponds in Qeshm Island. The Kharboz Aub-anbar dates back to Sassanid period. This pond is located near the ancient town of Kharboz in the southwestern part of Qeshm and was built in the midway of this town to the town of Pakhazar [31].

Though the information and the documents in hand to carefully investigate the materials, tools, style, and the methods of building Aub-anbars in the past is not sufficient for present-day curious researchers, relying on the remains and the prevalence of Qanats, Aub-anbars, and dams in Sassanid and Ashkanid eras does provide good evidence of significant progress in construction techniques in the past. Being aware that the architecture of those types of facilities had not been different from the factors and principles practiced in other types of structures, and still paying attention to the reality that not long ago and even at present, what we have is a heritage from the past; then, we can make cautious and intelligent guess about the past. For example, presently, the construction of a doom upon a square cross-section in relatively small sizes is possible; thus, we can imagine that in the old days, constructing cubic Aub-anbars with four-part roofs had been possible too. Also, having figured out the technique of dooms' structure on simple baldachins, the construction of reservoirs with six or eight cross-sections had been possible, although those types of reservoirs, because of the difficulties in isolation and extra cost, were probably not built very often. Maxim Siroux thinks that the cylindrical or pillar-type Aub-anbars came to Iran through the captive soldiers from the Mediterranean region in Sassanid period [25, 26].

One of the oldest roofed Aub-anbars is Azodi Aub-anbar which was built by the orders of Azod-al-dawla Deilami (911–958) in Estakhr, Fars Province. From that time to the Safavid period (1486–1714), where the construction of Aub-anbars became common place by kings and benefactors, we can find Aub-anbars whose dates of construction are clear and some are mentioned below [27, 32]:

1. An Aub-anbar near Robat-e Tahmalaj with a dome roof of 17 m in diameter and 8 m heigh. It belongs to the 9th and 10th centuries (823–973).
2. In the city of Marv, near the shrine of Mohammad Ibn-e Zayd, an Aub-anbar was discovered which belongs to 11th and 12th centuries. It has

a cylindrical reservoir with 6.1 m in diameter aired through two small windows. It has no covering, but it might not have had it from the day of construction.

3. Seyed Esmail Aub-anbar in Tehran which was built in the early 11th century. It was once repaired and renovated in Safavid period and then again in Qajar time [33].
4. Aub-anbar in Yazd which dates back to the time of Shoja King (1344–1365) is still functional and people use it.
5. Zam Zam pool in Gazorgah in Herat which seems to be built by order of Shahrokh, son of Taimour (1386–1429).
6. Jennok Aub-anbar in the alley behind Jam-e Mosque in Yazd and belongs to 1457.

Except some other old Aub-anbars, the other ones in Iran belong to 16th and 17th centuries and Safavid, Zandieh, and Qajar time. In some of the travel accounts of tourists in Iran, there have been references to Aub-anbars in different parts of the country.

The famous geographer, Moghaddasi, from the 10th century, writes about Aub-anbars [34]: "This desert (Kavir) is like a sea: you can travel from any direction, of course if you know the way and do not keep out of sight the water ponds and their domed roofs."

Also, Naser Khosrow in his travel account mentions [35]: "In this desert, there is a domed roof every 2 Parsangs[2] (or Farsangs) for the rainwater to be collected and the water is not salty. These domes are also used as signs for the travelers not to be lost in the desert and they are meant to rest in for a while if they are tired or it is very hot or cold."

In 1614, Garcia de Silva Figueroa was dispatched to the Court of King Abbas Kabir when he was 44 years old. His trip to India and Iran took 10 years of which 2 years and 7 days he spent in Iran. During his stay, he visited Lar, Shiraz, Isfahan, Kashan, Qum, and Ghazvin and tens of other small towns and villages [36]. Figueroa, in three cases in his travel account, refers to the Aub-anbars of Lar city and believes that the water of those Aub-anbars is the best in the world. He states: "In the house of many rich people, there are private Aub-anbars. There are many public Aub-anbars in towns and in the plain; even in the location where white tents were established, you could find Aub-anbars storing the best water of the world."

Pietro Della Valle, born in April 1586 in Rome, and after marriage, with his wife and some attendants and servants, he traveled to Iran on 4th of January

---

[2]1 Parsang = 6.24 km.

1617 [37]. In his travel account, Pietro refers to Aub-anbars and says: "On 6th day of our trip, we could cover 4 Parsangs and it was in the third Parsang of the trip that we arrived in an Aub-anbar which was built under the ground and had several steps. Since no spring was around and the land near the Aub-anbar was unsuitable for farming, therefore, it was there just for the convenience of the travelers."

Jean Baptiste Tavernier, French tourist, is one of the great explorers and experts to Iran in (17th century) who was very keen in knowing and noticing the details of everything he observed so he would report what he had seen honestly and clearly [38]. Between 1632 and 1668 AD, he had six trips to East and visited Iran for nine times. In his travel account, he writes about the Aub-anbar of Lar: "In Lar and around, there are many big Aub-anbars. It is necessary to have as many Aub-anbars as possible, because there is no rain for two or three years. When it rains, on the first day of raining, the windows of Aub-anbars are shut not to let any water get into them because water running on the surface of the ground is not clean enough. Then on the second day, water is led into them."

Maxim Siroux writes about Aub-anbars as [26] "The quantity of water in Qanats varies through the year. It is much from April to July. Some times Qanats become dry due to various reasons. So, Aub-anbars, which are filled in spring, are essential. In some cases, these Aub-anbars are so huge that they can procure the water needs of the house for 3–4 years."

Frederick Charles Richards, a member of Royal Academy of Painters and Engravers in England, made a trip to Iran in the early 20th century. His trip coincided with the rein of King Reza Pahlavi. He comments on the Aub-anbars of Yazd as [39] "The most interesting and the most magnificent architectures of Yazd, is not its mosques; rather its Aub-anbars . . . . These Aub-anbars with their Baudgeers which are used to keep water cool are praised by foreigners. Water in them stays surprisingly cool . . . . The water for these Aub-anbars is procured by 60–70 Qanats, running through the desert and located between the mountains around and the Aub-anbars."

References

[1] Moghtader, M. R. (1982). *Ab-anbar: Conservation de l'eau Sur le Plateau Iranian*, UNESCO, Paris, Decembre 1982.
[2] Varjavand, P. (1987). *Iranian Architecture, Islamic Period*, attempted by M. Y. Kiani. Jihad-e Daneshgahi, Tehran.

 [3] Peernia, M. K. (1987). *Iranian Architecture*, attempted by M. Y. Kiani.Jihad-e Daneshgahi, Tehran.
 [4] Dehghani, A. R. (2006). Consideration of history and process of delivery of water by Cisterns. *J. Air Cond. Ref.* 21, 4–11.
 [5] Dehghani A. R., et al. (2009). The holiness of water in history, legends, culture and civilization of Iran, in *Abstracts paper of the 2nd National Conference of Linguistics, Inscriptions and Texts*, Feb. 21 to 23.
 [6] *Holy Koran*. (2005), translated by E. Ghomshee. Masjed-e Jamkaran, Qum.
 [7] Dehghani, A. R. (2009). *Water in the Plain of Iran: Qanat, Cistern and Ice-house*. Yazda, Tehran, Iran.
 [8] Goublou, H. (2009). *Qanats (A Technique to get to Water),* translated by Dr. M. H. Papoli Yazdi and A. Sarvghad Moghadam Papoli.
 [9] Dehkhoda, A. A. (1998). *Loghat Nameh*, Vol. 3, 2nd Edn. Tehran University, Tehran.
[10] http://www.cloob.com./club/article/show/clubname/irancivil/articleid/86 064
[11] Hasib al-Karaji, A. M. H. (1994). *The Excavation of the Hidden Waters*, translated by H. Khadivjam, Tehran: The Research Center for Humanities and Cultural Studies.
[12] Labbaf Khaniki, R. A. (2004). *Gonabad: The Origin of Hidden Epics*. Tehran: Cultural Heritage Organization Center.
[13] Yazdani, E., and Godarznia, E. (2008). "The Construction of Indigenous Water Dams," in *The 1st Conference on Indigenous Technologies of Iran*, Sharif University of Technology, Tehran, Iran, May.
[14] Ibn-e Balkhi, *Farsnameh*, an attempt by G. Le Strange and R. A. Nicholson, Tehran: Donyai-e Ketab, 1984.
[15] Mubarak Bukhari, S., and Sari Oughli, K. E. (2004). Anis *al-talibin va'uddat al-salikin*, translated by T. Hashem Poursobhani, Tehran: Association of Works and Cultural Distinguished people, pp. 220–221.
[16] Chardin, J. (1993). *Travel Account of Chardin*, translated by E. Yaghmaee, Tehran: Toos.
[17] Mokhlesi, M. A. (1995). *The Old Ice-houses, Forgotten Masterpieces in Architecture*, A collection of Articles from the Congress on the History of Architecture and Urbanism of Iran, Vol. 2. The Cultural Heritage Organization of Iran, 690–695.
[18] Papoli Yazdi, M. H., and Labbaf Khaniki, R. (1999). "Ice-house and the production of artificial ice," in *2nd Regional Conference on the Change of Climate, Meteorological Organization of Iran*.

[19] Alizadeh Gohary, N., and Latifi, M. (2006). "Ice-house, a Masterpiece in the Heart of Kavir," in *Scientific Conference on the Architect of Kavir*, Islamic Azad University, Ardestan Branch.

[20] Abazari, J., and Tashakkori, B. (2008). "Qanat, Aub-anbar, and ice-house," in *A Project on Theoretical Architecture*, Department of Architecture, Islamic Azad University, Taft Branch, 2008.

[21] Bahadori, M. N., and Dehghani, A. R. (2010). *Natural and Traditional Ice Making in Iran*. Yazda, Tehran, Iran.

[22] Ghobadian, V. (2008). *Climatic Analysis of the Traditional Iranian Building*. Tehran University, Tehran.

[23] Safinejad, J. (1984). *The Fundamental of Human Geography with Reference to Human Geography of Iran*. Tehran University, Tehran.

[24] Karimian, N. (1999). *Culture of Water and Traditional Irrigation*. A Publication of National Committee of Irrigation and Drainage, Tehran.

[25] Abedini, M., Saeedi M., and Alemzadeh, H. (1995). *Aub-anbar, Islamic Great Encyclopedia*, 2nd Edn, Vol. 1, attemted by M. K. Mousavi Bojnourdi. Tehran: The Center for the Islamic Great Encyclopedia.

[26] Siroux, M. (1949). *Caravansérails d'Iran et petites constructions routiers*. MIFAO, Cairo.

[27] Farshad, M. (1983). *The History of Engineering in Iran*, 2nd Edn. Bonyad-e Nishabour and Goiesh, Tehran, 252–256.

[28] The Palestinian Encyclopedia. (1995). *Islamic Great Encyclopedia*, attempted by M. K. Mousavi Bojnourdi, Vol. 1, 2nd Edn. The Center for the Islamic Great Encyclopedia, Tehran.

[29] Abbasi, D. (2001). *The Architecture of Water, the Old Knowledge of Iranian People in Confronting Water Shortage*, Iran Newspaper, 1935, 19.

[30] Moradi, N. (1997). *Water Saving, Management of Water and Rain*. *Salehin-e Rousta J.* 136(15), 2–14.

[31] Javaheri, P., and Javaheri, M. (1999). *A Solution for Water in Fars History*, Tehran: The Iranian National Water Treasure & National Committee of Irrigation and Drainage, Vol. 2.

[32] Massarat, H., and Dehghani, A. R. (2008). "The Study of Typology of Aub-anbars in Iran," in *Proceedings of the 1st Iranian of Indigenous Technologies Conference*, Sharif University of Technology, Tehran, Iran, May.

[33] Blouk Bashi, A. (1992). *Aub-anbar in the Children & Youth Culture-book*, under the supervision of T. Mirhadi and E. Jahan Shahi. (Tehran:

The Company for the Provision and Publication of Children and Youth Culture-book).

[34] Ahmad-e Moghaddasi, A. M. (1982). *Ahsan al-Taghasim fi Marefat al-Aghalim*, translated by A. N. Monzavi, Tehran: Iran Company of Authors and Translators.

[35] Khosrow, N. (1976). *Travel Account of Naser Khosrow,* attempted by M. Dabir Siaghi. National Works Association, Tehran.

[36] Garcia de Silva Fig.

[37] UEROA. (1984). *Travel Account*, translated by G. Samiee. Nou, Tehran.

[38] Della Valle, P. (2002). *Travel Account (the section related to Iran)*, translated by S. H. Shafa. Scientific and Cultural Publication Company, Tehran 110.

[39] Tavernier, J. B. (1957). *Travel Account*, translated by A. Khajeh Nouri, corrected by H. Shirazi, Isfahan: Sanaee Library, 3rd Edn., 676.

[40] Richards, F. C. (1964). *Travel Account*, translated by M. Saba. Bongah-e Tarjomeh and Nashr-e Ketab, Tehran.

3

The Architecture of Aub-Anbars

The architecture of Aub-anbars in different areas have been under great influence of the local architecture, so that it has become a part of the architectural identity of the area. Experts and specialists of architecture, urbanism, history, and structure have classified Aub-anbars based on criterions such as: application, the rate of water collection, the arrangement of the steps and other spaces, the accessibility method to water, the shape of the reservoir and the dome-coverage, the method of the construction, the method and the type of water collection, their peculiar position in towns and neighborhoods, size, the method of airing and cooling, and the kind of decorations and the materials used in them. The ideas of experts in classifying Aub-anbars have features in common and differences as well which we will discuss in this chapter.

3.1 Typologies of Aub-Anbars in Iran

3.1.1 Function

3.1.1.1 Public Aub-anbars
Among the Aub-anbars in Iran, only a limited number had private, home application. Most of Aub-anbars, like the ones built in cities, villages, deserts, etc., had a public application and were used by everybody. Generally, public Aub-anbars were eye-catching, benevolent, big buildings, which benefactors, rulers, kings, and the rich would construct for the satisfaction of God and people, and the expenses for the construction, maintenance, water collection, repair, and renovations were provided by donors or government.

3.1.1.2 Private Aub-anbars
The Aub-anbars which were built in towns and villages in a private land or for the exploitation by the landlords and their family members were private Aub-anbars. Those Aub-anbars were mostly built in the houses and sometimes

Figure 3.1   An Aub-anbar under the yard in Gonbad-e Kavous [4].

in the private gardens. A number of desert Aub-anbars were used by a certain group of farmers in certain areas, which could be included among private ones. House Aub-anbars were of three types: under the yard, under the building, or beside the house wells [1, 2].

1. Under-the-yard Aub-anbars: These Aub-anbars were of rectangular shape and with a level or cradle-type ceiling. To get to the water, a bucket was used through a window or a manual pump pulled up the water [3]. An example of this kind of Aub-anbar in Gonbad-e Kavous is shown in Figure 3.1;

2. Under-the-building Aub-anbars: In this type of Aub-anbars, there was a space called 'tap-foot', which was at the same level as the reservoir and a brass-made faucet through which the water was accessible. Faucets were placed a little above the bottom of the reservoir to avoid the mud and other sediments to get into it;

3. Beside-the-house-well Aub-anbars: In the house of some rich people, water reservoirs were placed higher than the building or beside the water well. In this way, while procuring the water needed by the family the small pool in the yard and fountains could have the water as well to give a beautiful view of the yard. Some of those Aub-anbars had the capacity of 3–4-year consumption [3, 5].

Figure 3.2   The entrance of Mosalla Aub-anbar in Naein which is outside the garden.

3.1.1.3 Private-public Aub-anbars

Some of the Aub-anbars were used both as private and as public, and Aub-anbars of Dowlat Abad in Yazd and Mosalla in Naein are among this type. Mosalla Aub-anbar in Naein located in a private garden has two accessing canals, one for the residents of the house and the other for public (Figure 3.2).

3.1.2 Application

3.1.2.1 City Aub-anbars

Undoubtedly city Aub-anbars are the best examples of the glory of architecture and construction in Iran. It is in this type of Aub-anbar that the architect has demonstrated his knowledge, skill, and experience in architectural art such as the construction of Baudgeer, domed roof, stair case, etc. all in one structure.

City Aub-anbars were usually built in the center of the neighborhoods and beside the religious, educational, welfare, and business places. The remaining ones from the old days have more capacity compared to the types of Aub-anbars that could procure the needs of the populated areas for months. The importance of these structures to the cities was so much that the materials for their construction were carefully chosen, and they were built with high quality. In addition to the essential parts of the structure, other portions such as big front yards, entrances, inscriptions, halls, wide stairs, tall Baudgeers, and also various decorations were added to the building [3, 6].

Figure 3.3   A picture of 'Six-Baudgeer Aub-anbar' in Yazd.

One of those famous city Aub-anbars in Yazd is 'Six-Baudgeer Aub-anbar' (Figure 3.3).

City Aub-anbars were built beside the residential areas, charity buildings, and on the paths and main passages so that more people could take advantage of them. The bigger the neighborhood or the city, the more the number of Aub-anbars that would be built. The number of Aub-anbars in Yazd is over one hundred, which are different in the number of Baudgeers, the size of the reservoir, and the number of stairs. In fact, it is difficult to find two similar Aub-anbars in this city.

3.1.2.2 Village Aub-anbars

There is not much difference between city and village Aub-anbars. Actually, the most important feature of village Aub-anbars is its simplicity of shape and structure without decorations (Figure 3.4). This type of reservoirs was often in the central part of the village and in cylindrical shape. The materials used in them were usually what were found in the area. The priority of the locations was near Qanats or the path of the rivers and flood water. Since villages were not much populated, there were only two or three Aub-anbars built in the center or near it. The Zein Abad Village on the way from Yazd to Taft has just one Aub-anbar on the margin line (Figure 3.5) [7].

The Makers of the Aub-anbars tried their most to build them beside mosques, religious centers, and bazaar so that more people could use them. In some villages, there are also en route, farm, and house Aub-anbars.

Figure 3.4 Aub-anbar in Sharif Abad village, near Yazd City.

Figure 3.5 An Aub-anbar beside Zein Abad Village near the route of Yazd to Taft.

3.1.2.3 Desert Aub-anbar

The Aub-anbars built in deserts, farm lands, or plains to provide the water needed by trespassers, farmers, and sometimes for animals were called 'desert Aub-anbars'. Those types of reservoirs were in shape of pools with muddy or brick ceilings. There was some space for resting all around the pool for people. To reach the water in those reservoirs, one could use a bowl or even his hand. Many of these pools had stairs to be used as the level of the water went down. For this reason in Yazd and some other towns these pools were called 'Manual Aub-anbars'. The use of the term does not imply that they were watered through Qanats, springs, or wells.

In the *Islamic Great Encyclopedia*, Abedini states that the main reason for building those Aub-anbars in arid deserts was to procure water for the livestock and writes [3]: "the reservoir of them was generally in square shape and the walls were about 2 m high on the ground with ceilings". But, Kardovani asserts that these desert Aub-anbars would collect water from the rain running on the ground [8]. In a general classification, the Aub-anbars on the routes of caravans could be included among desert Aub-anbars since they were fed with the rainwater.

In the plains around the towns of Yazd province, there are many of these muddy reservoirs scattered in farms. Most of them are built close to each other and they have the name of the person who built them and even some had endowed properties [7]. In some of those Aub-anbars, a shelter was also built for resting and praying. In some, a room was also built to keep the live stock in [7]. Desert Aub-anbars are not deep. Professor Mohammad Ghaderi distinguishes 'farm Aub-anbars' from 'desert Aub-anbars' in that the former was used for the farming purposes whereas the latter were not. He states that: "the water in farm Aub-anbars was mostly used for farming purposes" [9].

3.1.2.4 House Aub-anbar (private or personal)

House Aub-anbars were mostly built in the towns located at the margin of Kavir or in rich people's house. In the description of them one reads: "Houses of Yazd each had a well and a reservoir. The wells were rather deep and people had to bring up the water by using a bucket or well-wheel and pour it into that reservoir" [10].

Those small Aub-anbars were built in the houses and were weekly filled with water from the Qanats, springs, rivers and sometimes rain. They were only used by the residents of the houses. When pipe laying system began to appear in towns, these reservoirs were forgotten; although in the villages without water pipes, they are still in use.

3.1.2.5 Mountain Aub-anbars

Aub-anbars of this type were built in the mountains (Figures 3.6 and 3.7). They were built out of stones using chisels and hammers (Figure 3.8). Some of them have arched ceilings. In some cases, a part of mountain formed a portion of the reservoir and the other portion was built by bricks and other

Figure 3.6   A view of forty ponds of Eij, Fars [11].

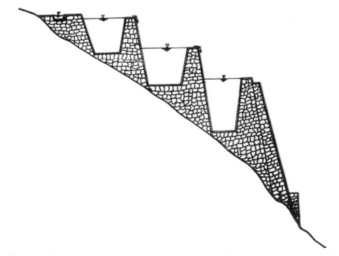

Figure 3.7   A sketches of ponds in the foothill of the mountain [11].

Figure 3.8   Water entrance of one of the ponds and the stone coverage of other ponds [11].

materials, like the Huge Aub-anbars of Damghan Mountains [3]. It is obvious that the rainfall would have filled those reservoirs.

3.1.2.6 Citadel Aub-anbar

It was in Ibn-e Balkhi's *Fars Nameh* that, for the first time, there was a mention of citadel Aub-anbars in Estakhr citadel, Semiran citadel, Abadeh citadel, and Ram Ravan citadel [12]. The most famous citadel Aub-anbar was the one built in the 12th century in Estakhryar citadel by the order of Azod-al-dawla Dailami. Its roof was held on twenty columns and it could provide water for one thousand warriors for one year in the time of battle [13]. It is quoted from Ibn-e Balkhi that: "In that land, there is no other citadel older than that and in that valley-shape fort, there was a piece of deep land which the rain gathered in from one side and went out to desert. Azod-al-dawla built a dam on the desert side and made a pool using stones, mortar, and plaster with a staircase of 17 steps" [14].

The commonest way of procuring water for Aub-anbars was the rain and then Qanat; nevertheless, in some places such as Hesam al-molk citadel in

Kermanshah, the water was provided by a well using an ox and winch to pull up the water from [15]. These types of Aub-anbars are found in all the citadels in Yazd. They are simple in structure and are often in the shape of covered pools. Their reservoirs were square-shaped wells, rather small and deep built in the central part of the citadel. Some of them were linked to the other buildings of the citadel in such a peculiar way that the water rain of those buildings was led to the Aub-anbar too [3].

3.1.2.7 Caravanserai Aub-anbar

Those Aub-anbars were built inside the caravanserais and were very much like the citadel Aub-anbars. Their water resources varied according to their locations by rain, Qanat, spring, or well. They were covered pools built in the yard of caravanserai with a reservoir appropriate to the population of that place. The roof was quite level and used for loading goods [3, 5].

The most difficult task for the en route caravanserais is procuring water. In some cases, water had to be procured from a very long distance. A stream of 25 km, Houz-e Soltan area, is the evidence for the claim [16]. One of the vital issues in desert caravanserais, particularly in Kavir and its marginal areas, was water procurement and its storage. The existence of an Aub-anbar inside or beside a building shows its importance in those days. The number of buildings with Aub-anbar outside or beside is more than the ones with Aub-anbar inside. In some cases, water was provided through Qanats from a very long distance [17]. Another way of furnishing water was the roofs of caravanserais. The roofs were built in such a way that the entire rainwater would run into the reservoir.

3.1.2.8 En route Aub-anbars

Some of the desert and caravanserai Aub-anbars may be known as En route Aub-anbars, but often the structures which researchers know by this name are built beside caravanserais (Figure 3.9). Some of them were also built along the roads and the paths between towns and villages [18]. In Figure 3.10, a sketch of En route Aub-anbar including the reservoir, the doom upon the reservoir and the stairs to reach the water, plus the covered space as the resting area for the trespassers is shown.

Benefactors would provide a cool and comfortable place for the tired and thirsty travelers of Kavir, where they could drink cool water and rest for a while. Even in some Aub-anbars, rooms were provided for people to rest and say their prayers [3].

Figure 3.9   Kolar Aub-anbar in Maybod in Yazd.

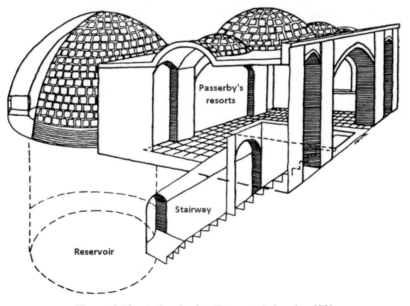

Figure 3.10   A sketch of an En route Aub-anbar [20].

In the book, *Under the Sky of Kavir,* Aub-anbars are described as [19]: "Along the road in Kavir, every few kilometers, there is a ditch dug like a pool with a domed roof on it. There are two holes on two sides of the doom where the running rainwater on the plain could get into the pool from. The doom had also been a guiding sign in Kavir for the passengers as well. The important function of the dooms is to keep the water away from the sun shine. Therefore, it can keep the water cool and not to evaporate. The Two holes on the sides would keep an air stream to flow through and air the water in the reservoir".

3.1.3 Procuring Water

3.1.3.1 River (flooding)

Although some of the village and desert Aub-anbars were fed with river water, but most of the en route Aub-anbars were watered by spring floods of the nearby rivers. For this reason, a curved levee was created in the riverbed, which led the water to the canal connected to the reservoir [21]. Saadi, the famous Iranian poet, has referred to the spring floods filling the reservoirs, in one of his couplets:

> Swiftly decided the old man to get back
> Filled Aub-anbar with spring flooding attack

The Aub-anbars which are filled by permanent or seasonal rivers are usually big and are constructed with bricks in a circular shape and domed roof and are equipped with Baudgeer. The rivers would mostly water the public Aub-anbars. Nevertheless, in some towns and villages of arid areas of the country where there was some noticeable surface water, water was conducted to the houses as well [8].

3.1.3.2 Rainwater

The history of Aub-anbars that were watered through collected rain from roofs and yards of the house goes back to ancient European civilizations. Those people had used this structure to procure their drinking water and other needs. The construction date and use of those types of Aub-anbars relates to the prehistoric time in the ruins of NAS Palace, the center of Gunmetal Era, in the island of Corte in France [22, 23].

In his book, Estakhri talks about a few en route pools (Aub-anbars) in Yazd which had been filled with rainwater [24]. Naser Khosrow in the 11th century in Zeil Garmeh outskirts of Khour and Biabanak refers to those Aub-anbars and writes: "and in this desert every two Farsang, dooms have been constructed which the rain is collected in" [25].

Kardovani has divided the rain-filling Aub-anbars into two groups [8]. The first groups are those Aub-anbars in which the rain collected from the roofs to fill them. The second groups are those which collect the rainwater from the surface of the ground (Desert Aub-anbars). The first group was often for

Figure 3.11   The collection of water through drain pipes, from the roof to the reservoirs, in Gonbad-e Kavous.

houses, but some en route and caravanserai Aub-anbars also took advantage of the same system of collecting water. In some parts in the south of Iran like Bushehr province where people were deprived of drinking water, the same method was used for watering reservoirs. The roofs of the houses were built in such a way that the rain through a special drain pipe would stream down the reservoir under the ground. The capacity of the reservoir depended on the amount of the rain, the area of the roof, and the size of the reservoir. These Aub-anbars have water faucet and stairs which are devised outside the reservoir. In some arid parts of the world, such Aub-anbars have been observed and, because of easy and free access to drinking water, are among the most useful of house Aub-anbars. An example of collecting water from the roofs and storing it in an Aub-anbar in the city of Gonbad-e Kavous is shown in Figure 3.11.

According to Kardovani, the second type of Aub-anbars was desert type and built in deserts. They were not watered by people; they were filled with the rainwater in a natural way. Those Aub-anbars were normally constructed beside caravanserais or en route the old routes and in a shape of river-water Aub-anbars; i.e., big and domed shape with bricks. They are seen in some villages in Iran, but because mud and dirt are not in a condition to be used, generally, these Aub-anbars were built in the places which had an appropriate soil with vast area for the collection of water (Figure 3.12) [8].

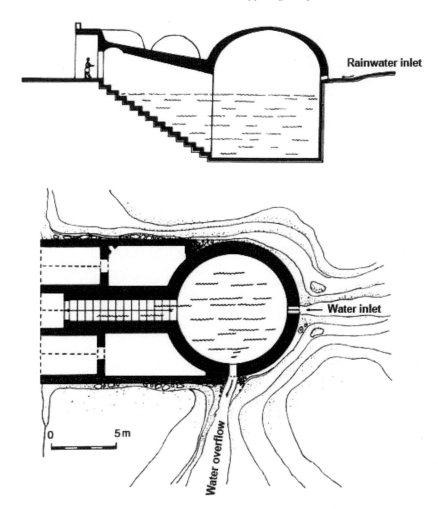

Rainwater inlet

Water inlet

Water overflow

0   5 m

Figure 3.12   A sketch of the plan and schematic cut view of Mohammad Ghasem reservoir in Tabas [27].

In this regard, Mostafavi also mentions: "establishment of ponds and rainwater reservoirs had been a common undertaking throughout the southern coasts of Iran and other places of Fars. There had been several connected water reservoirs in the outskirt of the mountains in Derayej, Estahbanat, since many years ago. In *Word Book*, the name of these sources has been mentioned, Sahrige" [26]. Figure 3.13 shows people of some village in rural areas of Minab, taking water by clay jars. Rainwater would gather in the natural ditches.

Figure 3.13   Taking collected rainwater from natural ditches in a village in rural areas of Minab [28].

Manual

The architecture of these Aub-anbars is in such a way that they are not fed with the rain or river water; rather by man through the transfer of the well water, Qanat, or springs from another location. In some cases, water had to run for hundreds of kilometers to get to the reservoir. A great number of city, village, and plain Aub-anbars, particularly in the arid areas of Kavir, were watered this way. The exploitation of Qanat water was done in two ways. In the first method, the access to the Qanat water was by building a public or private water-terminal and in the second method by leading the water to the reservoir of Aub-anbar [17].

3.1.4 Accessing Water

3.1.4.1 Covered reservoir

Most of the city, village, en route, and even house Aub-anbars have covered reservoir. These reservoirs are covered with domed roof or level surface (more house Aub-anbars). They are very suitable in terms of hygiene and their water is chilled and aired by Baudgeers or some windows on the roof. The materials used in them are mud-bricks and mortar. There is a staircase beside the reservoir, which leads to the faucet location. The faucet mounted in the

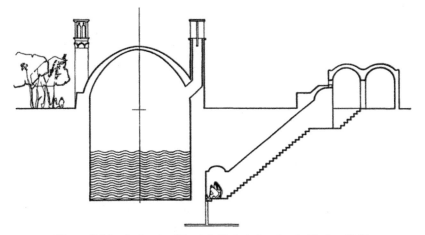

Figure 3.14   A sketch of Panjeh Shah Aub-anbar in Kashan [30].

lowest portion of the reservoir is the only way to access water. Figure 3.14 shows a cross-section of Panjeh Shah Aub-anbar in Kashan with staircase, faucet area, and faucet.

3.1.4.2 Open reservoirs

This kind of Aub-anbar, also called pool or pool-Aub-anbar and according to Baghban Zadeh: "open Aub-anbar is the one from which water can be taken by hand, a bucket or a bowl" [9]. They have two different structures. In some of them, there is no faucet and the staircase is inside the reservoir [16]. In the other one, water is pulled up by rope, pulley, and bucket; as the Chahar-Sou Houz in Herat, which is described by Eslami Nodooshan [29]: "It is an open Aub-anbar known as Chahar-Sou pool and it dates back to 17th century. People still take water from it". There are some Aub-anbars in Iran similar to Chahar-Sou in Tabas and Mohammadieh in Naein (Figures 3.15–3.17) [16].

The stair-like structure of these Aub-anbars includes a water reservoir which is connected to the alley or the square by a stair from the floor. The water is pulled directly up. As the water level goes down because of consumption, more steps are needed to access the water. The other type of open reservoirs are the ones which were located alongside the alleys, and had a small reservoir. The opening of the Aub-anbar was in a corner of the alley and the water was pulled up manually by a bucket or a pulley. Other devices were used as well such as jar and goatskin.

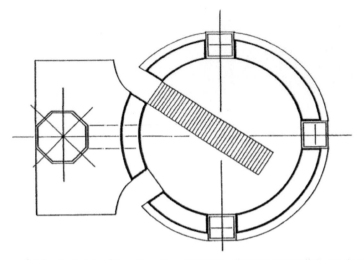

Figure 3.15   A sketch of the plan of an Aub-anbar in Mohammadieh, Naein [30].

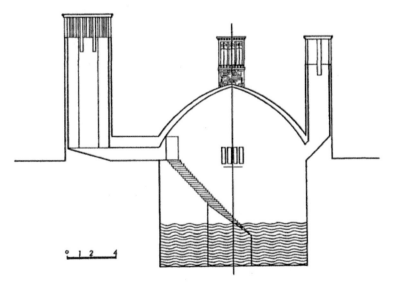

Figure 3.16   A sketch of the schematic cut view of an Aub-anbar in Mohammdieh, Naein [30].

In some parts of Iran, water was taken from Aub-anbars by other devices like Eshkelak in Tabas. Eshkelak was a piece of wood (Figure 3.18) which was somehow connected to the jar and thrown into the reservoir to be filled, then it was pulled out.

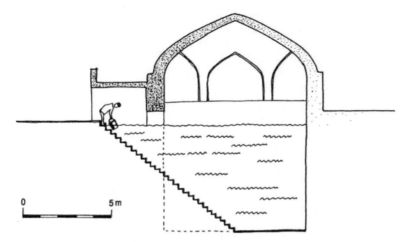

Figure 3.17 A sketch of the way the water was taken from a pool (An Aub-anbar whose stairs were in the water) [27].

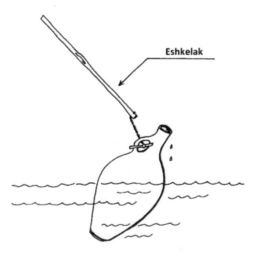

Figure 3.18 A sketch of a device for taking water from the pool (An Aub-anbar without stairs) [27].

3.1.5 Ventilation (Airing)

3.1.5.1 Baudgeer or Khishkhan

In some Aub-anbars, there is a Baudgeer or Khishkhan which is used to ventilate the air above the water and keep it cool (Figure 3.3).

Figure 3.19   A view of an Aub-anbar in Deh Namak, Semnan province.

3.1.5.2 Openings (holes)

The Aub-anbars which have Openings, lack Baudgeers or Khishkhan. To ventilate the air and keep the water cool, some windows are made on the roof or the body of the reservoir. In northern-east part of Iran, there are Aub-anbars which have a dome or a circular staircase the top of which is covered with a lantern-like structure to ventilate the system (Figure 3.19).

3.1.6 Materials

3.1.6.1 Stone-made

There are some Aub-anbars in some corners of Iran built with stone [31]. Figure 3.20 shows one of them, en route Tehran-Garmsar.

3.1.6.2 Mud-brick-made

There are Aub-anbars scattered in Iran built with mud-bricks. Even, in some, the reservoir and the Baudgeer were built from the same material. In some of them, in order to strengthen the resistance of the reservoir, it was covered with a layer of bricks or clay and straw mixture.

3.1.6.3 Brick-made

Majority of Aub-anbars in different parts of Iran were made with bricks; even some of the en route ones were made with bricks from top to the bottom. The

Figure 3.20 A stone Aub-anbar en route Tehran-Garmsar.

size of the bricks used in them is 20×20 cm. To add to the age of the stairs, they were made of the bricks provided at higher temperatures. These bricks were as hard as stone.

3.1.7 Appearance

Considering the kind of roof, Aub-anbars may be divided into two types: circular and flat.

Circular: They are built into the ground and cost much less. The reservoirs are whether cylindrical or rectangular. Their capacity varies from 20 to 30 m^3 in small ones and 1000–3000 m^3 in big ones.

Flat: The reservoirs of such Aub-anbars are long rectangular with the roof held on the columns inside the reservoir.

Aub-anbars in the south and southern coast of Iran are divided into the following types in terms of their covering:

Long reservoirs: the length is 2–3 times its width with one entrance and one exit (Figure 3.21).

Circular reservoirs: Circular Aub-anbars have a cylindrical roof and reservoir. Figures 3.22–3.24 show the plan, cross-section, and the view of Sheikh Yousef circular Aub-anbar in Bastak town in Hormozgan province.

Mixed reservoir: This type of reservoir is a combination of long and circular reservoirs. First, a circular reservoir is built in the center and, then, some long reservoirs are added to its sides. The entire structure looks like a star (Figure 3.25) [26].

Figure 3.21 A pictures of two long pools in Dejegan, Hormozgan province.

Figure 3.22 A picture of Sheikh Yousef circular reservoir in Bastak, Hormozgan province [32].

3.2 Parts of Aub-Anbar

3.2.1 Water Reservoir

The reservoir is the place where water is stored. It is the main part of the structure. The entire or a part of the Aub-anbar is built under the ground surface. Locating reservoir down the ground has always been for three reasons [31]:

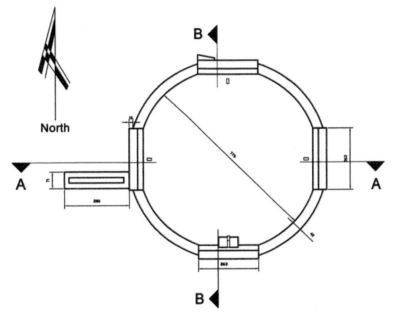

Figure 3.23 A plan of Sheikh Yousef circular reservoir in Bastak, Hormozgan province [32].

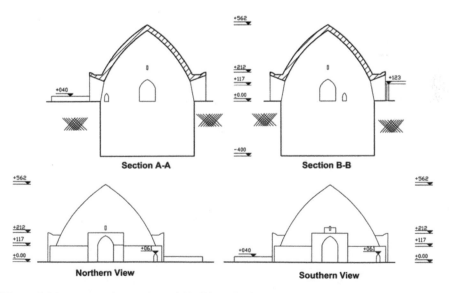

Figure 3.24 Schematic cut view of Sheikh Yousef circular reservoir in Bastak, Hormozgan province [32].

Figure 3.25   An Aub-anbar in Bandar-e Lengeh. Since its reservoirs are like a plus sign, it is called Chahar Berke (Four-reservoir or star-reservoir) [31].

1. If the reservoir was built on the ground, the pressure of the water to the walls would destroy it. To alleviate this pressure, there was a need for thick supporting walls, which would cost a lot. But if the reservoir is built under the ground, the surrounding soil of the reservoir resists the pressure exerted to the walls by the water inside. They are also much more resistant against earth quakes. Aub-anbars and other under-ground structures have been remained relatively undamaged or slightly damaged through all the earthquakes that occurred in the past.
2. When the reservoir is under the ground, the water from the stream or Qanat may be naturally and easily transferred to it and there is no need for extra energy to do that.
3. The more the depth of the reservoir under the ground, the less is the variation of the temperature and beyond 6.5 m; the temperature is equal to the average annual temperature of the surface ground. Therefore, water does not freeze in winter (as the water in the well) and it is cool and refreshing for drinking in summer.

To have a better understanding of reservoirs, there is a need for some classifications. Probably, the most important criterion in classification of reservoirs is the location as city, village, en route, desert, etc., Aub-anbars. In other words, the reservoirs with circular or cylindrical structures were used in cities, villages, and en route. The ones with square plan, in square and rectangular cubic, or 8-sided shapes were built in farms and deserts and in some instances in towns and villages. The variety of reservoirs in different places was due to financial issues, the structure strength, the volume of the

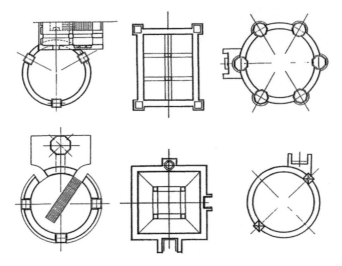

Figure 3.26 Plans of the various Aub-anbars in Iran [30].

required water, and the importance of the structure. In towns and villages, since Aub-anbars played a vital role in daily life of the people, the strength of the structure and the volume of the water were very important. For this reason, they were built in cylindrical shapes; but, in places such as farms or en route, where the strength and the water volume were of less significance, they were built in square shapes, rectangular, or 8-sided shapes. In Figure 3.26, plans of various shapes of Aub-anbars are shown.

The diameter of the cylindrical reservoirs varies from 5 to 20 m and is covered with domed roofs. The dome was almost as high as the diameter. This type of reservoirs has a capacity of up to 3000 m^3. In a number of huge Aub-anbars where the construction of cylindrical and big domes was impossible, columned-square reservoirs with the ceiling and domes were used. A good example of this type of Aub-anbars is seen in Hosseinieh Zavareh (Figure 3.27) [33].

Square and rectangular cubic reservoirs may be constructed not only in small sizes but also in big sizes too. In big ones, to hold up the ceiling, columns and supports were applied inside the reservoir. The Aub-anbars of Hormoz Island and Seyed Esmaeil in Tehran are two examples of this type [34]. The cover of the most reservoirs was domed (hemispherical), conic; staired-conic (Figure 3.28), hollowed semi-cylindrical (in Persia called Ahang [35]) and flat (Figure 3.29). Most often, the cover of the reservoirs is painted white to absorb

Figure 3.27 A view of a columned reservoir in Hosseinieh Zavareh.

Figure 3.28 The brick-cover of an Aub-anbar in Garmsar.

less heat (Figure 3.30). Sometimes, beautiful decorations were drawn by bricks on the domes of Aub-anbars, like Rostam Giev Aub-anbar in Yazd with several cypresses as the symbol of magnanimity of Iranians. The decoration is more visible when the rain makes it wet.

To construct small domes in rural areas and villages, layered and flat stones are used and for the covering of the big- and medium-sized domes, bricks are used. The flat roofs, apart from their use in the columned big reservoirs, are also used in the Aub-anbars, which are built in conjunction with mosques, schools, caravanserai, warehouses, or similar structures in a complex. On the roofs, structures such as mosques, chambers, and the like were constructed [34].

Figure 3.29 The flat-cover of the reservoir of Khayyam Aub-anbar in Ghazvin.

Figure 3.30 A view of an Aub-anbar en route Lamard-Ashkenan with a white domed roof [11].

Normally, there is one reservoir for each Aub-anbar; but, in some rare cases, the number could be two (Aub-anbars of Hossein Abad and Asr Abad, near Yazd, which both have two reservoirs with seven Baudgeers and Kish's Aub-anbar with two reservoirs and five Baudgeers (Figure 3.31)). Rostam Giev Aub-anbar in Yazd has one main and two side reservoirs (Figure 3.32) [3, 7].

Figure 3.31   A picture of an Aub-anbar in Kish Island with two reservoirs and five Baudgeers.

Figure 3.32   A view of Rostam Giev Aub-anbar in Yazd.

Almost in all Aub-anbars, there are two bricks mounted in a zigzag fashion, on the inner wall, which are protruded for approximately 10 cm from the wall and with about 90 cm distance from each other. These stairs which are called footsteps in local language are extended from the entrance window of the water (or air) to the bottom of the reservoir, used for dredging, cleaning the sludge, repair, or other maintenance activities. Some ropes were also connected to the ceiling or the sides of the reservoir to be used for going up and down the reservoir. Those ropes kept the workers safe while working in the tank. The reservoirs of the desert Aub-anbars have a separate area with stairs and tap-foot.

In the roof of Aub-anbar, one, two, or three holes were made. One for the entering of the water and the other two which were located 0.5–1 m below the first hole, called over-flow, were used for the exiting of the wood chips and trash while filling the reservoir. The holes were completely sealed after filling to prevent the animals from getting into the reservoir.

3.2.2 Baudgeer and Ventilator

Baudgeers are the structures which belong to centuries ago and they were used to cool the air in buildings. They were built in some parts of Middle East and Egypt, where the climate was hot and dry or hot and wet. They were also used in Aub-anbars to keep the flow of air and cool off the water [36]. Baudgeers were very tall, big, and glorious in Kavir and arid areas of Iran. They had different shapes like cubic rectangular, hexagons, or octagons prisms and cylindrical.

At the top part of Baudgeers are several openings in various shapes and thin walls divide the trunk of the Baudgeers into different parts. This way, even the slightest air would be transferred to the surface of the water in the reservoir and push the warm and humid air out through the windows made in the roof, or dome or the Baudgeer itself [3, 37].

Based on the number of Baudgeers, Aub-anbars are classified into: mono-, bi-, tri-, tetra-, penta-, hexa- (Figure 3.3), and heptagonal Baudgeers (Figure 3.33). Most Aub-anbars have two or four Baudgeers. The architecture of Baudgeers varies from one to the other. In Figure 3.34, the only example of two tetragonal Baudgeers whose openings are honeycombed is seen.

It is located near the Zoroastrians' dungeon in the city of Yazd. The Baudgeers of the Aub-anbars in the towns of Ardakan and Maybod are normally one-sided and face the north wind (Figure 3.9). The height of Baudgeers is from 5 to 12 m.

Figure 3.33   Asr Abad Aub-anbar with seven Baudgeers.

Figure 3.34   An Aub-anbar with two honeycombed Baudgeers beside the Zoroastrians' dungeon.

In Aub-anbars with no Baudgeer, a vent is made at the top part of the domed roof and a small ventilator is built, called Khishkhan, a Baudgeer-like structure but in small and short to establish air flow (Figure 3.28), in the east of Khorasan province, in Taybad area, [33] in Hormozgan and Fars provinces

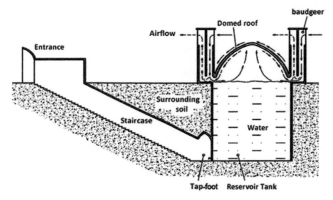

Figure 3.35 A cross-section of an Aub-anbar and the normal airflow in the Baudgeer and the vent on the roof.

in order to ventilate the air in the reservoirs, and four windows were made at the four corners of the dome.

In Figure 3.35, the direction of the airflow and its circulation in the reservoir is shown, in an arid area. As it is seen, the air gets into the Aub-anbar space through the wind-facing canals and after passing over the water, it gets out through the wind-backing canals and the window at the top of the domed roof.

3.2.3 Staircase and Tap-Foot

Normally, beside the reservoir and in the middle of the entrance (entering door), the staircase was built which made it possible to reach the tap-foot and take the water [33]. The location of the staircase and the reservoir depended on the size and the condition of the piece of land used to build Aub-anbar. These two parts can be in different conditions with respect to each other:

1. Reservoir and the staircase are along the same line (Figure 3.36);
2. Reservoir and the staircase are parallel (Figure 3.37);
3. Reservoir and the staircase are perpendicular (Figure 3.38);
4. A combination of parallel and perpendicular stairs (Figure 3.39).

In some cases, like the staircase of Ghal-e Nodooshan Aub-anbar in Yazd, it is irregular and almost spiral. In general, the staircase was built in a straight line from the entrance to the tap-foot to have enough light for the tap-foot [38].

To have enough light for the long path of the staircase, two methods were used. In the first, big, vertical canals were used in different parts of the staircase roof and usually above the landings and on the floor. To prevent crash into

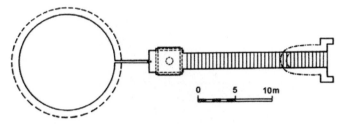

Figure 3.36   A sketch of Chahar-Sou Aub-anbar in Tabas, where the reservoir and the staircase are along the same line [27].

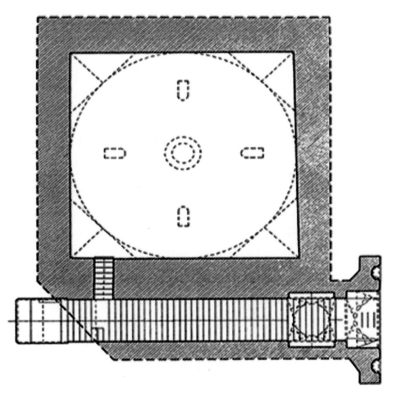

Figure 3.37   A sketch of an Aub-anbar in Ghazvin, where the reservoir and the staircase are parallel.

the canals, a shelter was built (Figure 3.40). In the second method, a part of the path was kept open with no roof. In this type of Aub-anbars, the width of the staircase was wider than the roofed part. The widening of the staircase top was not just for the light (Figure 3.41).

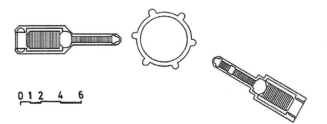

Figure 3.38   A sketch of an Aub-anbar with six baudgeers in Yazd, where the reservoir and the staircase are perpendicular.

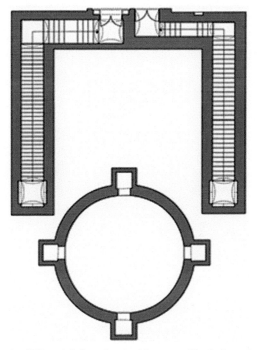

Figure 3.39   A sketch of Kiyani Aub-anbar in Aharestan, Yazd city, where the reservoir and the staircase are both parallel and perpendicular.

Some of the big Aub-anbars have two different separate staircases in order to ease the trespassing. In some Aub-anbars, like the Aub-anbar of Biouk Alley in Yazd, one of the staircases had been for the Muslims and the other one for the Zoroastrians (in some parts of Iran, the cultural and religious issues would influence the architecture of Aub-anbar). Sometimes one staircase was for

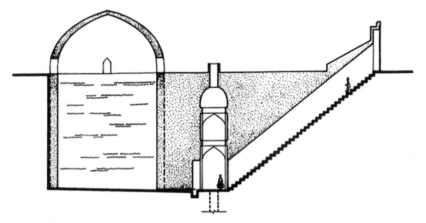

Figure 3.40   A sketch of Chahar-Sou Aub-anbar in Tabas, Yazd province, Iran [27].

Figure 3.41   A picture of six-Baudgeer Aub-anbar in Yazd.

the public and the other one for private use, such as the Mosalla Aub-anbar (Figure 3.42) and Dowlat Abad Aub-anbar in Yazd [3, 39].

The number of stairs varied with the height of the reservoirs. Regarding the great need to store the water, the height would reach at 15–16 m. Shah Abol Ghassem Aub-anbar has got 76 stairs. Sometimes the staircases were too

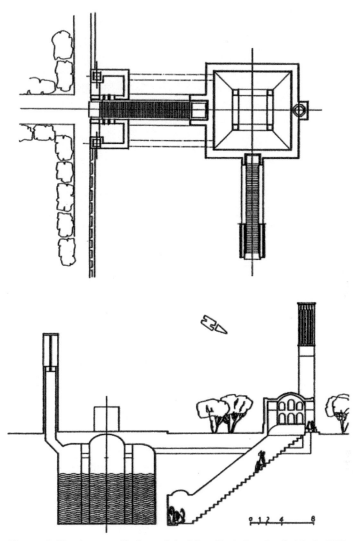

Figure 3.42 An overall view of the Mosalla Aub-anbar in Naein [30].

steep but wide enough to ease the passing of two persons simultaneously. The height of the stairs was, normally between 25 and 30 cm and from the floor of the stair between 30 and 50 cm. Every few stairs, there was a flat area for the water takers to rest. The width of the stairs was the same in all the path or it would become narrower every few stairs. Stairs were built with or without landings. Usually, in the landings, one or two small or big rooms were built for resting or selling goods [38]. In some Aub-anbars, there are two landings. The covering of the staircase was in double-layered with the help of arches and Roomies (this type of structure is used for the wide openings because of its strength. In this structure, which is done by bricks or mud and bricks, the structure seems as a fence from the vertical and facing views; and as a complete screen in a horizontal view). The style of bricking the staircase, the roof, and the walls was in various shapes, while very plain.

Some staircases have two steep and flat ceilings with one to few meters empty space between them. This way, the heavy load of the ceiling was avoided, and also, the space between the ceilings acted as an insulator against the humidity from the upper to lower ceiling. Moreover, the steep of the staircase ceiling would be lesser and the ceiling of the staircase to be the same as the floor of the passageway [40].

Tap-foot is a small room at the same level as the reservoir bottom and in semi-octagon or square shape. It often has some platform on two sides for sitting. There is a water-tap made of brass in the tap-foot and a small ditch with a stone or metal cover to transfer the sewage to underground canal. The tap is usually situated one meter higher than the bottom of the reservoir to prevent the exiting of the sediments in the reservoir through the tap [3, 7, 30].

Water-taps are often made of whole brass and decorated with various designs. This shows the artistic taste and the symbolic value that the Iranian artists would link it to the life and their own works (Figure 3.43) [30].

Regarding the number of residents in the area and to avoid any waste of time, there were, sometimes up to three water-taps in the tap-foot. In some Aub-anbars to facilitate the taking of the water, the taps were mounted at different height. As the level of the water in the reservoir fell down, the lower taps were used [34]. An interesting point which is an indicative of the Iranian creativity and sense of architectural insight is that a bottleneck structure was

Figure 3.43   A picture of two water-taps made of brass for Aub-anbars.

built behind the tap inside the reservoir to reduce the pressure of the water and by which the tap would be kept safe.

Another point worth mentioning is that a small ventilation window was also made on the roof or tap-foot to ventilate the air and provide some light too. In Mohammadian Aub-anbar in Kashan, in order to cool the adjacent house, a canal was extended from the tap-foot to that house [33].

3.2.4 Entrance

Entrance is the decorative part of the structure through which one can enter an Aub-anbar and its stairway. It is the most eye-catching part as well. Usually,

Figure 3.44 The inscription of Massoum Khani Aub-anbar in Naein, made of marble.

entrances are made of one big middle arch, two side-pillars, two smaller arches at both sides of the big one, and the inscription on the forehead of the entrance. There are semi-arches at the entrances which give a glorious view to it. All parts of the entrance are covered with various kinds of decorations and designs. Colorful tiles (7-color tiles) are usually used in the inscription. Sometimes the middle portion inscriptions are engraved in marbles [34].

The size of the entrances are different and depend on the size of the Aub-anbars, the position in the texture of the whole structure, extend of the passage, the front square and the importance of the adjacent buildings. Normally the entrances are taller and bulkier compared to the adjacent buildings. On both sides of the entrance, there are wide stone-platforms to connect the structure to the open space and a place for the people to get together.

Generally, the entrance of the Aub-anbars was placed at the beginning of the stairs; but, in those, where the staircase was partly open, the entrance was built in the beginning of the covered part of the staircase. In fact, the open part acted as a small passage [40].

In front of the entrances, considering the location of Aub-anbars, there has been an open space, if possible, to add to the view of the entrance. For this reason, Aub-anbars of any neighborhood were built beside the mosque and the square to create the most significant part of the location. Even the ones in Bazaar had an open space in front. In some cases, to give more prominence and beauty to the entrances in covered paths, the roof of bazaars and passages were kept open and roofless [34].

The investigation of inscriptions at the entrances is very important to know more about the cultural back ground and religious beliefs of the people of the time [34]. Most public Aub-anbars have inscriptions including the name of the sponsor, builder, calligrapher, engraver, tile setter, poet, and occasionally the name of the king or the ruler of the region, the text of the endowment or the poems about the sponsor (Figure 3.44). The important content of the inscriptions is the poems about the religious figures of Islam such as Imam Ali (peace be upon him), Imam Hossein (peace be upon him), and others as well. In some, Aub-anbars are referred to as pools and ponds, and the value of them has been compared to the sea or Zam Zam spring like the inscriptions of Malek Bazaar (Tabriziha Bazaar) in Kashan where the last couplet is as follows [42]:

> Said, the one, took water from this pond,
> Zam Zam is embarrassed of the flowing pond

The innovation introduced by the architect of the traditional structure, an example of which is seen in the Six-Baudgeer Aub-anbar in Yazd, is that he has not built any entrance at the open part of the staircase, but by making dents in the pillars, has created a vertical space (Figure 3.41).

There is not a complete and definitive example of the entrances prior to the Safavid era; however, there are many entrances remaining from the post-safavid period, the inscriptions of which are quite helpful in their identification [34]. In Figures 3.45–3.48, the entrances of different Aub-anbars are shown.

3.2.5 Details of the Façade (View)

The builders of some of the Aub-anbars, by creating exquisite designs in the façade of the structure, apart from expressing their thoughts and feelings, had other ideas in mind too, i.e., those elements had precise functions [40]. Usually, that part of the structure that was exposed to the destructive elements such as direct strokes to the body and the rainwater penetrating into the mud-bricks were made of tiles and bricks in beautiful, innovative designs.

The desert Aub-anbars were made of mud, small in size, a Baudgeer-like composition, a circular reservoir, and a simple roof covering the staircase and possibly a small inscription which bore the name of the constructor and the donor.

The commonest village-city Aub-anbars were constructed using bricks laid in flower-designs with colorful gems among them. The seven-color tiled and decorated inscriptions were used in some Aub-anbars like Jennok in Yazd

Figure 3.45 The entrance of Mosalla Aub-anbar in Naein.

Figure 3.46   The entrance of Boresteh (Boresseh) Aub-anbar in Taft.

Figure 3.47   The entrance of an Aub-anbar in Naragh.

Figure 3.48 The entrance of an Aub-anbar in the town of Kooh Payeh.

and Boresteh (Boresseh) in Taft (Figures 3.49 and 3.50). There are some other decorations in Aub-anbars such as decorative arches, plasterwork, decorative pillars, special bricks, etc.

3.3 The Method of Construction and the Materials Used in Aub-Anbars

To build the Aub-anbar, this engineering masterpiece, responsible to hold and passively cool the precious jewel of the hot and dry and hot and wet regions of Kavir, i.e., water, had to go through several stages mentioned below [40]:

1. Materials of the structure;

Figure 3.49 The inscription of Jennok Aub-anbar in Yazd, mounted at the entrance [30].

Figure 3.50   The inscription of Boresteh (Boresseh) Aub-anbar, mounted at the entrance.

2. Coating of the reservoir;
3. Method of construction.

It should be mentioned that the location of the Aub-anbar was chosen in a place where the land was firm and could bear the heavy walls of the reservoir and its ceiling, particularly when it was filled with water.

3.3.1 The Materials of the Structure

Since Aub-anbar was directly in touch with water and wetness, Aub-anbar was built by the kind of materials which were resistant against destructive elements and wet. The main materials used in the construction of Aub-anbars were stone, lime-mortar, and mortar. The unbaked and baked bricks were used for the back and the body of Aub-anbar, and special lemon-color bricks, known as Aub-anbar-brick were just used for the reservoir. This kind of brick was used in the construction of the Chinese wall, here was used in the coverage of the reservoir, body, and staircase. Of course, in mountainous areas or where the stone was available being more cost effective than bricks, pieces of stone were used for the walls and the domed roof. The mortar used in the construction of the structure had often a kind of lime in its composition. Generally, the floor of Aub-anbar was covered with a lime-mortar and in some Aub-anbars like khajeh and Rig in Yazd and Ganj Ali khan in Kerman, to strengthen the floor and prevent any change in its form; a layer of lead was used too. The lead could also keep the water cold. The mortar used in the Chinese brick was a combination of ash, sand, and lime and the mortar of the back wall of the reservoir was a mixture of mud and lime.

The surface of the internal walls of the reservoir was isolated with a mixture of clay, lime, and ash. Lime was mixed with clay with a ratio of 4:6 and the mixture was diluted with water to produce a dense paste. The paste was stirred with a long piece of wood and then was put aside for two days to harden. After enough time, the ashes and other ingredients were mixed to get their density back. Next, the provided coating was spread over the internal surface of the walls and then the sands were added to make the coating quite hard. When the drops of water, which were known as the sweats of the mortar, appeared on the walls, Aub-anbar was ready to be filled with water. The design of the Aub-anbars is in such a way that it stores the water in low temperature in summer. To do this, it takes advantage of the temperature variations in desert areas and the isolating characteristics of the land [30].

In addition to bricks, mud-bricks were also used for the outer surface of the walls and stones, at times, for the tap-foot and staircase. The outer surface of the domed roof and the walls of the Aub-anbars in the south of the country were coated with a special mud with whitish-gray color. This mud was extremely resistant against wetness and, because of its color, it would not let the water of the reservoir to become too warm [3].

3.3.2 The Method of Construction

The construction of Aub-anbar can be divided into two stages of digging-excavation and construction. To carry the soil out of the ditch after excavation of the covered part of the land, a tunnel was dug in the ground. The wall bricking was done after mortar establishment of the floor of the reservoir. The walls were as thick as three bricks and there was about one meter of mortar-establishment behind the wall. The last thing to be done was to roof the reservoir. In doing this, no cast was used and the roof had varying thicknesses in its curvature.

Building the Chinese wall and roofing the staircase were other stages of construction. The excavation was done regarding the weight of the bricks and the mortar. The roof was built in arch-shape or roman-shape each with its own specific technique. At the end, the roof was covered with mortar.

There were other points to be noted in the construction of Aub-anbars, including the digging of a well to drain the polluted water inside the reservoir and directing it to a dry Qanat, the building of a place for the water-tap which was done with a lot of care, and the construction of the Baudgeer and connecting it to the coverage [40].

In some places, to save money, a special kind of Aub-anbar was built known as 'Rikhteh-ey' [40]. The method of the construction was like this: the ground was dug as much as the half of the thickness of the wall of the Aub-anbar, to get to the favorite depth. Then, filled that place with lime-mortar and waited for two weeks to let it sag. Next, the center of that place was dug for the main reservoir and mortar was spread over the walls and the body. Now, the reservoir was ready to be watered. Sometimes, the mortar used for the floor of the reservoir was mixed with stone-pieces to give it more strength. The strength of the adjacent wall of the staircase led to the tap-foot was very important, because this part of the structure was not surrounded with the soil, as the other parts. Therefore, it was built thicker and more resistant than the other parts.

References

[1] Massarat, H. (2010). *Aub-anbars of Yazd*. Yazda, Tehran.
[2] Massarat, H., and Dehghani, A. R. (2008). "The Study of Typology of Aub-anbars in Iran," in *Proceedings of the 1st Iranian of Indigenous Technologies Conference*, Sharif University of Technology, Tehran, Iran.
[3] Abedini, M., Saeedi, M., and Alemzadeh, H. (1995). *Aub-anbar, Islamic Great Encyclopedia*, attemted by M. K. Mousavi Bojnourdi, 2nd Edition, Vol. 1. (Tehran: The Center for the Islamic Great Encyclopedia).
[4] Eskandarnejad, A. (2008). *The Qualitative and Quantitative Study of the Water in the Aub-anbars of Gonbad-e Kavous*, BS Thesis, Department of Environmental Health Engineering, Islamic Azad University, Medical Sciences Branch, Tehran, Iran.
[5] Siroux, M. (1949). *Caravansérails d'Iran et petites constructions routiers*, MIFAO, Cairo.
[6] Pour Moghaddasi, H., et al. (2003). Chemical and Microbial Investigation of Village Aub-anbars in Golestan Province. *Water Wastewater J. Isfahan*, 45, 26–31.
[7] Memarian, G. H. (1993). *A Survey of the Architecture of Aub-anbars in Yazd*. Iran University of Science and Technology, Tehran, Iran).
[8] Kardovani, P. (1989). *Resources and Issues of Water in Iran*. Tehran University, Tehran.
[9] *An Interview with Mr. M. Ghaderi and Mr. A. A. Baghban Zadeh* by H. Massarat.

[10] Besharat, H. (2007). *Yazd, My City*, Yazd: Andishmandan-e Yazd, 2nd Revision.

[11] Javaheri, P., and Javaheri, M. (1999). *A Solution for Water in Fars History*, Vol. 2. The Iranian National Water Treasure & National Committee of Irrigation and Drainage, Tehran.

[12] Ibn-e Balkhi (1984). *Farsnameh*, an attempt by G. Le Strange and R. A. Nicholson. Donyai-e Ketab, Tehran.

[13] Le Strange, G. (1958). *Historical Geography of the Eastern Lands' Rulers,* translated by M. Erfan.: Bongah-e Tarjomeh and Nashr-e Ketab, Tehran.

[14] Mostoufi, H. (1915). *Nezhat al-Gholoob*, attempted by G. Le Strange and Liden.

[15] Golzari, M. (1978). *Kermanshahan, Kurdistan*. National Works Association, Tehran.

[16] Varjavand, P. (1987). *Aub-anbar, in Tashayo-e Encyclopedia*, Vol. 1. under A. Javadi and et al. Islamic Institution of Taher, Tehran.

[17] Eyvazian, S. (1995). "Trend of the Formation of Out of Town Caravanserais," in *Collection of Articles in the Conference on the History of Architecture and Urbanizing in Iran*. Vol. 1 (Tehran: The Organization of Cultural Heritage of Iran).

[18] Afshar, E. (1975). *Mementos of Yazd*, Vols. 1 and 2. National Works Association, Tehran.

[19] Mohajer, A. A. (1961). *Under the Sky of Kavir*. Habibi, Tehran.

[20] Poya, A. (1992). *The Ancient View of Maybod*. Islamic Azad University, Maybod Branch, Yazd.

[21] Mostoufi Bafghi, M. M. (1961). *Jam-e Mofidi*, Vol. 3, attempted by E. Afshar. Asadi, Tehran.

[22] Sotoudeh, M. (1977). *Aub-anbar*, Danesh Nameh Iran and Islam, Vol. 1, under E. Yar Shater. Bongah-e Tarjomeh and Nashr-e Ketab, Tehran).

[23] Ahmad Jeihani, A. (1998). *Ashkal al-Alam*, translated by A. A. Kateb. Astan-e Ghods-e Razavi, Mashhad.

[24] Estakhri, A. E. (1961). *Masalek and Mamalek*, attempted by E. Afshar. Bongah-e Tarjomeh and Nashr-e Ketab, Tehran.

[25] Khosrow, N. (1976). *Travel Account of Naser Khosrow,* attempted by M. Dabir Siaghi. National Works Association, Tehran.

[26] Nour Bakhsh, H. (2002). Water Ponds in Persian Gulf. *Farhang-e Mardam J.*, 1(2), 78–91.

[27] Danesh Doust, Y. (1997). *Tabas, a City from the Past; the Historical Buildings of Tabas*. The Organization of Cultural Heritage of the Country and Soroush, Tehran.

[28] Bakhtiari, M. (2001). *A Detailed Guidance of Iran: Hormozgan Province*, 2nd Edn. Geographic and Cartographic Institute of Iran, Tehran.

[29] Eslami Nodooshan, M. A. (1978). *The Whistle of Simourgh*, 4th Edn. Toos, Tehran.

[30] Moghtader, M. R. (1982). *Ab-anbar: Conservation de 'eau Sur le Plateau Iranian*, UNESCO, Paris.

[31] Ghobadian, V. (2008). *Climatic Analysis of the Traditional Iranian Building*. Tehran University, Tehran.

[32] Poudat, S. (2005). *The Documentation Report of Shiekh Yousef Pond*. The Organization of Cultural Heritage and Tourism, Hormozgan Province.

[33] Ayatollah Zadeh Shirazi, B. (1970). The Aub-anbars in the Margin of Kavier, *Iran. Archeol. Art J.* 5, 30–36.

[34] Varjavand, P. (1987). *Iranian Architecture, Islamic Period*, attempted by M. Y. Kiani. Jihad-e Daneshgahi, Tehran.

[35] Shams, S. (2009). *Traditional Glossary of Iranian Architecture: Effects of Iranian Architecture Art.* Elm-o Danesh and Nou Avaran Pars University, Tehran.

[36] Bahadori, M. N., and Dehghani-sanij, A. R., and Sayigh, A. (2014). *Wind Towers: Architecture, Climate and Sustainability*. Springer International Publishing, Berlin.

[37] Dehghani, A. R., Mazidi, M., and Eslahchi, A. (2007). *A Study of the Operation of Wind Towers, Dome Roofs, Caps and Cisterns*, 2007 World Renewable Energy Conference-Pacific Rim Region, Publication in Proceeding, Taipei, Taiwan, 29 October–30 November 1, 224.

[38] Dehghani, A. R. (2006). Study of History and Evolution of Wind Towers, This Beauty and Nice Phenomena and Desirable Traditional air Conditioning System. *J Air Cond. Ref.* 19, 4–13.

[39] Dehghani, A. R. (2006). *Analytical and Numerical Investigation of Heat Transfer in Cisterns,* MS Thesis, Department of Mechanical Engineering, Islamic Azad University, Science and Research Branch, Tehran.

[40] Mortazavi, M., and Bagheri, M. (2008). "Technology of Water Procurement in Kavier: Qanat and Aub-anbar," in *Proceedings of the 1st Iranian of Indigenous Technologies Conference, Sharif University of Technology*, Tehran, Iran.

[41] Tavassoli, M. (1981). *Urban Structure and Architecture in the Hot Arid Zone of Iran*. Payam and Peyvand-e No, Tehran, 82–92.

[42] Farrokh Yar, H. (2007). *Aub-anbar, a Forgotten Memento*. Helm, Qum.

4

Health and the Quality of Water in Aub-Anbars

4.1 Characteristics and Various Contaminations of Water

Naturally, water runs on the surface or under the ground and contains various elements in the form of solution, suspension, and colloid. These elements, based on the way of consumption, can be classified as proper, improper, or dangerous. Therefore, investigation before consumption is essential.

It is extremely important to carry out physical, chemo-physical, chemical, biological, and radio-logical tests in order to identify the existing elements in the water, to determine its quality for drinking, industrial, and agricultural use. The analysis of the water can help detect the probable problems of the water before use and, in turn, pave the road for finding the solutions to solve them. The kind of the water consumed by a great number of people and some industries like food or pharmaceutical industries must be controlled steadily to keep it safe from microbial and organic pollutants. In addition to tests and the analysis of water, we need other information such as the origin of the water, the way it gets out of the ground, the method of transfer, the way of distribution, etc. In suspicious cases, one-time test is not enough and the final results cannot be determined. To find out about the microbial pollution, more tests should be repeated [1, 2]. Here, some of the characteristics of water that play an important role in the quality of drinking water would be referred to.

4.1.1 Drinking Water

Drinking water of communities should not only be enough, but also be suitable from health point of view. Per capita consumption of drinking water is approximately 1–3 l. Generally speaking, the factors affecting the quality of water include:

(1) Kind and the amount of chemical compounds;
(2) Bacteriological features.

4.1.2 Sources of Drinking Water

The main water resources are underground resources, surface water, and streams. Water wells, springs, and Qanats are considered underground water. Dams' water, rivers, and seas are considered surface resources. Although, they are abundant, but they might not be suitable for drinking. It seems those resources will also be somehow treated and used as drinking water in future. Streams are formed by the rain and snow to some extent. A part of earth's water exists in the atmosphere in the form of vapor and still some part is stored in ice form in natural glaciers of the poles.

4.1.3 Properties of Drinking Water

To specify the properties of drinking water is not an easy undertaking, because the water is not a synthetic product. One cannot specify constant properties for it. Thus, some activities should be done on the water to make it suitable and safe for drinking. Those activities known as the refining process make the water drinkable in terms of its physical, chemical, and microbial content.

1. *Appearance*: Recognition of the physical properties of potable water does not guarantee that it is suitable for drinking. It is just a guide to use it. Drinking water should be pleasant, clear, colorless, and odorless.
2. *Physical properties*: One of the important physical properties of water is its temperature. In limited volumes, water temperature is influenced by the ambient temperature; but, in big volumes, like lakes and seas, it influences the ambient temperature. Generally, water temperature can be used as a source of information to know its origin, volume, exit, penetration, and its depth under the ground. Drinking water in different locations has some salts, which give them different degrees of electrical resistance, which is another physical property of water. The identification of this feature in water piping networks and water currents can show its contamination with other substances. In fact, the degree of water resistance or its electrical conductivity can help clarify different facts about the water under consideration.
3. *Chemical and physiochemical properties*: In addition to above-mentioned features, water contains various minerals, which are needed

by the body of the human beings. Water passes through different regions of the land and, based on the type of the land, it passes through, might become contaminated, and cause diseases. Some substances are not dangerous on their own but in combination with other substances they can be dangerous. One of the important physiochemical properties of water is its pH. Generally, most potable waters have a pH of approximately 7; their pH can fluctuate between 6 and 8. Drinking water may contain some air, i.e. oxygen, nitrogen, and carbon dioxide, each of which having its own particular effect. Oxygen can biologically affect water, nitrogen can affect the lightness of the water, and carbon dioxide can affect the chemical balance of water. Underground water has less oxygen compared with surface water.

4. *Bacteriological properties*: Usually, most natural waters and deep-water wells in some certain cases contain micro-organisms that enter water through air and surface waters. The germs in the volume-unit of water vary considerably. Waters containing organic substances include more bacteria than the waters without them. Use of polluted water before filtering may cause some problems. Origin of pollution may be quite different in different cases. Nevertheless, they can be specified. The pollution can be microbial contamination in water-distributing networks, penetration of polluted water of towns and villages into the ground, or the surface sewage and its release into the ground, and even the penetration of polluted water from far distances. The major danger for the drinkable water to become polluted is the sewage systems. In biological tests, the Colibacill count is often a good index for the contamination of water by excrement. Among the bacteria of the digestive system, which is found in the water, is *Escherichia Coli*, considerably more than other Coliforms in the sewage systems of houses, because its origin is the excrement of humans. There are several forms of Coliforms, which possess different health properties and they can be determined in micro-biological tests. Most Coliforms are not found in natural waters and, if found, it means that water is polluted. *E. Coli* has excremental origin but it cannot say that the other bacteria of the Coliform group in the water do not. Coliforms, except *E. Coli*, are more resistant and their existence is a sign of old pollution which needs further tests. Coliforms count is the main part of biological tests of water, which can specify new pollutions [1, 2]. Among other microbes which can pollute the water, we can refer to *Streptococcus Faecalis* or Enterocoques, which are the bacteria of the digestive system and their number in excrement is less than that of

E. Coli. All groups of Enterocoques can be found in the digestive system. They do not proliferate in water. Their existence in water is the sign of new excremental pollution of water. Salmonellae can cause acute Gastroenteritis accompanied with diarrhea and the spasm of abdominal muscles along with fever, nausea, and vomiting. Salmonellae are usually found in polluted water like sewage systems, pools, basins, agricultural water, and floods.

4.2 Water Hygiene in Aub-Anbars

4.2.1 Filtration and Disinfection

Hygiene and procurement of safe water has always been the concern of authorities of the city and physicians many years ago in Iran. Safe water means safe and healthy community. Our ancestors would also investigate the quality of water carefully. There have been ways and procedures to specify the properties of the safe water and the dangers of polluted water as well. For example, in the book The *Extraction of the Hidden Waters* written by Hasib al-Karaji, it is stated that [3, 4]: "Whenever you see that the color of water has changed, you should know that it is not a good water and if you smell a bad odor, it is the sign of pollution and if you taste and find it unpleasant, it is not safe". Elsewhere in the book, he states that: "the most refreshing water is the rain and snow water which run through the sweet soil or pass through sand and pebbles with no water vegetation. The water which possesses other properties than the above mentioned, have been affected by the soil and the plants in its running path. Moss and alga can change the taste of water and produce salty, bitter, sulfuric, heavy water".

Regarding the chemical properties of water, al-Karaji says that [4, 5]: "In order to find out the superiority of two different waters with similar physical properties, weigh the same volume of them to see which is lighter. The lighter one is the better one".

For Aub-anbars, when it was full with the water, at once or after a short time (two or three days), after its contents would sink to the bottom, depending on the size of the Aub-anbar, salt-stones in three forms, solution, semi-stone, and stone, were thrown into the reservoir from the windows of Baudgeer or other windows on the dome or the walls. The salt-stone, for example, for Ghandehari Aub-anbar in Yazd was 60 kg in a year and for Dowlat Abad Aub-anbar it was 48 kg. The reason for the salt in the water was to prevent it from putrefaction. It is quoted from Jorjani that [6]: "The nature of water changes if it remains in the same place for a long time".

The money for the salt was provided either by people living in the neighborhood or by some donors whose names would be written in the plaque of the Aub-anbar [7]. In some cases, lime stone was used to disinfect the water in the Aub-anbar. When the salt dissolved in the water and the dust would settle, a layer of salt formed on the surface of the water, which prevented the water from decaying. People believed that if the salt layer tore or found a hole, water would go decay or as they used the expression 'the seal of water was broken' [8]. Therefore, after filling the Aub-anbar with water, all its holes were carefully sealed. Sometimes the layer of salt was broken by birds or falling of a thing into the Aub-anbar. In order to prevent the decay of the water, once again salt was thrown into the water and all children were advised not to throw anything into the Aub-anbar [9]. In some Aub-anbars, coal dust was added to the salt for the forming of the preventive layer on the water [10]. Although salt makes the water salty, but because of being static, the salty substances would sink and refreshing water was available [11].

Another cause for the decaying of water was that water ran through unclean streams, polluted and mixed with the remains of animals' excrement and the sewage of the houses, although they would sink to the bottom of the Aub-anbar. Sometimes, small animals entered into the reservoir and their dead bodies made the water decayed in which case, water in had to be disinfected.

In short, there were two physical and chemical filtrations carried out on the water. In physical filtration, all the small floating bits, after the water became still, would sink to the bottom of the Aub-anbar. This process quickened when clay and lime were added to water. In chemical filtration, salt was added to the water. Salt decomposed in the water and chloride was released and disinfected the water. In recent years, chloride has been used to disinfect the water in Aub-anbars [12].

Our ancestors, to prevent the stinking of water, had found a solution and it was placing pieces of coal or a bag of ashes into the water [13]. In some Aub-anbars in Iran, fish were placed in the water. The fish would eat the eggs of pests and save the water. In some places like Fars, consumers of water suffered from Filariasis. This disease was caused by the egg of a worm which hid itself under the skin of the body [14].

Since Aub-anbars were covered with roofs, they would become less polluted. The coverage not only prevented the evaporation of water, but also kept it away from contamination [15]. Comparing the uncovered Aub-anbars of the south of Iran with the covered ones in the north, experts believe the water in northern Aub-anbars had more clean and refreshing water [16]. In discharging the uncovered Aub-anbars or ponds, buckets were used which usually transferred the remains of the animals' feces, wandering around the ponds, into the water. This would pollute the water. The tests of uncovered Aub-anbars indicate that the number of Colibacill could reach over 1000/cm^3 of the water, while there should be no Fecal Coliform in the drinking water. Moreover, different types of, Khak-shee, larva, and worms are observed with unaided eyes in the water [17].

This has been mentioned about the epidemic of Piok disease in Iran [18]: "The pools had been the source of several diseases, particularly Piok. But in recent years, because of health care and adding chloride to the water of pools, Piok has entirely eradicated". In a book titled *Piok disease*, Sadid al-Saltaneh states that the disease is peculiar to Larestan and southern ports of Iran. It is a thin, long worm, which is created in the body of humans [18].

Another danger which the pools, coastal areas of Persian Gulf, and Aub-anbars are threatened by is the accessibility of Aub-anbars to everybody without any restriction. Staircase is inside the reservoir and people barefoot, with shoes on, or dirty hands and even dirty buckets or other containers take the water from the reservoir. The most important thing is that the windows of the reservoir are open through which dust and other polluting particles can get into the water in the reservoir. Furthermore, the sun through the windows shines onto the water and makes it go decay and helps microscopic organisms grow and multiply. Although water-pipes have been utilized for many years, but, still there are villages in the southern parts of Iran which have to use the pool-water [16].

In a systematic investigation of 114 rural Aub-anbars in the Golestan province, where the collected rainwater on the roofs was conducted to Aub-anbars, it was specified that the chemical elements including the ones effective in electrical conductivity, alkalify, hardness, chloride, nitrate, and manganese were at the acceptable level compared with standards of drinking water. But the density of some of the chemical substances such as iron (9%), lead (69%), and chrome (6%) were higher than the acceptable amount. In terms of microbial variables in 100 ml of the samples, the number of Coliforms in 56%, *E. Coli* in 32%, and *S. Faecalis* in 26% of the collected samples were higher than the standard. The quality of the water in a considerable number of Aub-anbars, in terms of chemical and microbial elements, was not suitable for drinking. The chemical and microbial contaminations of Aub-anbars might be because of the penetration of agricultural sewage, and human and animal waste into water and the use of river-water too [19, 20].

The researcher came to the conclusion that the rain as a natural phenomenon does not impose any chemical problem to human health, unless there is a severe contamination in the area. If fact, microorganisms get into although the rain washes little chemical elements in the atmosphere and brings them down, but, it is not one of the main sources of microbial contamination of water. Therefore, microorganisms in the water of Aub-anbars are through the area where water is collected or stored. Studies show that non-disinfected Aub-anbars are appropriate places for the disease-causing microorganisms. Some of these diseases include a digestive-system disease for which salmonellae are the cause: Ulcer which is caused by Helicobacter, lung infections like Ligonier, and blood infections caused by corny bacterium [19, 20]. The researcher believes that: "the collecting method of water is the main reason for the contamination of water. The water which is stored in Aub-anbars is collected whether from the roofs and yards when raining or through the river. The material which covers the roof of the houses, usually made of metal sheets or old ceramics, may affect the chemical property of the collected water and microbial contaminations in the yards can be carried to Aub-anbars with the water flow" [19, 20].

The researcher has presented some suggestions to optimize the situation of Aub-anbars as follows [19, 20]:

1. To clean up the water storage place once a year;
2. To filter the water physically before entering the Aub-anbar or before consumption and changing the filters according to the instruction of the producing factory;
3. To disinfect water by chloride and maintaining the remaining chloride at standard levels; using UV ray before consumption, and boiling the water.

Generally, Aub-anbars built in the Golestan province are rectangular in shape and hold between 25–35 m^3 water. The walls are usually made of cement with roofs. There is a small window on the top to discharge water through for consumption and other purposes. There is also a chimney to evacuate the air inside. All Aub-anbars in the area are built under the ground and the rainwater collected on the roofs of the houses is lead to these Aub-anbars [20].

Mohammadi and Shah Mansouri in a study [21] have investigated the quality of the water of Aub-anbars in town of Bandar-e Lengeh. In this study, 60 samples of the water in different Aub-anbars of 10 villages for 3 months were analyzed for their microbes including *E. Coli* and *S. Faecalis*. The tests were carried out according to multi-pipe fermentation procedure, differential tests and pour plate. The results of the tests have revealed that in 100 ml of the samples, microbial variables such as Coliform, *E. Coli*, and *S. Faecalis* exceed the maximum acceptable rate in 100% of collected samples. Human and animal feces and other factors are considered the most important microbial pollutions. The chloride residue of the water of all Aub-anbars was reported to be zero.

In a study by Dehghani et al. [22] they carried out the qualitative and quantitative investigations of Aub-anbars' water in the town of Birjand. In this study, 12 samples of water for the analysis of the chemical parameters and 9 samples for the microbial analysis were collected from different villages in two stages in April and July 2009. The results showed that the physical and chemical parameters including turbidity, pH, EC, total hardness, alkalis, chloride, sulfate, carbonate, bicarbonate, nitrite, nitrate, flour, calcium, magnesium, sodium, potassium, iron, manganese, ammoniac, and TDS in all Aub-anbars were at an acceptable level; compared with drinking-water standards. Only the density of some of the substances like Ferro was more in 16% of the samples and that of Flour was less than the standard in all samples. In terms of microbial parameters in 100 ml of the samples, the number of Coliforms in probable stage in 77% and in confirmation state in 33% of the samples was more than the standard. In 11% of the samples, Fecal Coliform was observed. The chemical quality of these Aub-anbars is at an acceptable level but not in terms of microbes. Supplying some provisions, it is possible to improve the condition of the water to a great extent.

Microbial, physical, and chemical tests were performed on the water of Aub-anbars in the villages of Sabzevar in Khorasan Razavi province [23]. In microbial tests, variables of Coliform and *E. Coli*, and, in physical and chemical tests, variables of pH, color, odor, turbidity, alkalinity, hardness, chloride, flour, sulfate, carbonate, bicarbonate, nitrite, and nitrate were examined. The tests show that all physical and chemical factors were within the standard limits except for one of the Aub-anbars in which the flour and chloride were beyond the limits. The probable reason for that could be the geophysical condition of the soil.

The results of the microbial tests indicated that the number of Coliforms in 50% of the samples and that of *E. Coli* in 40% of the samples were higher than standard and the main cause for that was the microbial pollution of the stored water.

Khani et al. have discussed various methods used for having healthy water since years. Those methods are briefly mentioned here [24, 25]:

1. Most of the Aub-anbars experienced total darkness. No water-plant and algae could grow in them because those plants need light to continue life;

2. All sinkable, suspended, and some salts, because of the gravity, would sink to the bottom of the reservoir. Also, salt and lime were added to the water which acted as the chloride used today to disinfect the water. Therefore, water filtering was carried out both physically and chemically;
3. Locating the reservoir inside the ground could prevent the freezing of the water in winter and provide the cool drinkable water for summer. One of the factors affecting the freshness of the water positively is the rate of dissolved oxygen in it. The more the oxygen content, the fresher the water;
4. The rate of dissolved gases in water depends on various factors such as temperature, pressure, and the like. Thus, the rate of dissolved oxygen in the water depends on the temperature. For example, the rate of dissolved oxygen in water at temperature $0°C$ is 14.62 gr/l and at $15°C$ is equal to 10.15 gr/l.

It should be noted that hygienic methods were not exercised in all Aub-anbars to the same extent; that is why the quality of water in different Aub-anbars was not the same.

The darkness of the Aub-anbar has been a preventive factor for the growth of various organisms in the water [16]. For the many reasons discussed above, Aub-anbars were filled in winter. Obviously, the cold and frozen season of winter would not let many disease-causing organisms to become active [26]. After all, physical filtering is one of the cleaning methods of the water.

In another study, Aub-anbars of the villages in Lar region were investigated for physical and chemical properties [27]. In April 2009, ten samples of water were collected from Aub-anbars of those villages and tested. The results indicate that the physical and chemical variables such as pH, temperature, alkalinity, nitrate, nitrite, turbidity, and floating oils in all of the Aub-anbars under investigation were at the appropriate level; nevertheless, some chemical variables such as electrical conductivity, total hardness, chloride, sulfate, and fluoride in some of the Aub-anbars were below the standard levels. The percent of the samples above the authorized level for each variable is at maximum 40% out of the total samples. For example, TDS, magnesium hardness, and magnesium ions in the samples were 40% higher than the acceptable rate. Regarding the fact that the Lar region does not receive enough rain during the year, the physical and chemical qualities of the water in Aub-anbars for some of the variables are not good enough; nevertheless, using some provisions can improve the physical and chemical condition of the water in those Aub-anbars.

Few centuries ago, al-Karaji has explained the ways which can be used to filter the water [4, 9]. He states: "if you want to turn heavy and salty water into a drinkable liquid, pour the water into a clay jar and let it leak from its bottom.

Or you may pass the water through layers of sand and pebbles. In Ardakan, people place bushes at the entering door to let physically filter the water".

In some areas, water was filtered in sandy pools and then sent into the reservoir [8, 28]. For the water to become clean and refresh, it is important that the salts and other sedimentaries to sink to the bottom of the reservoir by the force of gravity, particularly where the water is salty or semi-salty. In Kashan, adobes [28] and in other places, clay soil was used to desalt the water. That is why the water tap for discharging the water was mounted about 1 m above the bottom of the Aub-anbars to prevent the discharge of the sunken salts and other elements. Today, chemical tests are used to identify the salts in the water.

Many experts believe that the water in Aub-anbars in many parts of Iran, even Shiraz, is cooler and more refreshing than the pipe water in cities, particularly in terms of plaster. It has been mentioned that people of Shiraz would have taken a bucket of the Ghasr al-dashe Aub-anbars' water to their homes because that water lacked any plaster [16]. In other parts of Iran like the coastal areas of Persian Gulf, and Sea of Oman where the water was salty, bitter, and unpleasant, the water in ponds is much sweeter and refreshing than the water in pipes [18]; nevertheless, that water has its own problems as well. Thus, in some marginal areas of Kavir, despite the city piped water and refrigerators, people still use Aub-anbars as their supply of drinking water [16].

In some neighborhoods, people put the pieces of wood on fire and used its smoke to disinfectthe water in the reservoir. Smoke made a layer on the water and saved it from dangers. Also, barriers were placed in the canals of Baudgeers and the small windows leading to the reservoir to stop the birds or other animals from getting into it [29].

In the book, *The Old Tehran,* there is an *instruction* given for the filtering of water in the old days [30]:

1. When filling the Aub-anbar with water, fill it to the brim and let some water spill in order to get away with dirt and other unwanted items;
2. Put lime, coal, salt, and ashes into the water in Aub-anbar;
3. Do not use the tap to discharge the water in first few days. Take water from the top;
4. If there is a local Aub-anbar with no fresh water, just use its water for a few days and not more;
5. Pour vinegar and rose-water into the water if it has a bad odor;

6. If the water has lost its taste and gone bitter, first, boil it with vinegar, clay, and rose-water; then, let the substances to sink down. Now you can drink it, but in order to prevent the diseases which might be caused by this kind of water, have lots of garlic, onions, and vinegar or lettuce.

Of course, these directions could be effective as long as there was no microorganism in the water like Khak-shee, a small pin-like worm with a big head and a narrow end, and the like. If these worms were observed in the water, a thin piece of cloth was wrapped around the discharging tap to collect them in that piece of cloth. The cloth was cleaned every day. If the pollution was high and it was probable that some animals like a cat or a mouse has fallen into the water, people would use mirrors to reflect the lights of the sun into the Aub-anbar to find the dead animal and get it out of the water and finally, put a lot of lime, coal dust, and other disinfectants into the water [30].

To exploit sandy Aub-anbars was proposed as a comprehensive project by Hooman in 1970 (Figures 4.1 and 4.2) [31]. In this plan, passing of water

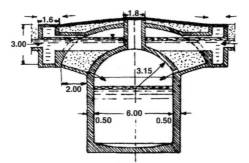

Figure 4.1   A sketch of sandy Aub-anbars [31].

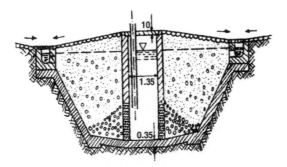

Figure 4.2   Another sketch of sandy Aub-anbars [31].

through sandy layers was the best way to filter the water. The advantages of this plan are briefly given below:

1. There is no need for ceiling and the cost of ceiling may be used to have bigger Aub-anbars and filling them with sands. For example, instead of having a circular cross-section with the diameter of 11.3 m, we can have a square cross-section with the side of 17.3 m and equal depth to hold three times water more;

2. Since this type of Aub-anbars has no ceiling, there is no need for them to have a specific shape; nevertheless, the floor and the walls should be constructed according to the technical and statistical calculations. They should be perfectly isolated not to let the rainwater leak from the bottom or salty underground or unsafe water to penetrate into it.

3. Since the shape of the reservoir is not important, any natural ditches en route the dried rivers may be used to save the digging cost and choose a size which is proportionate to the amount of the rainwater which might flow through that river.

4. All the sandy surface of the Aub-anbar can face the flood and be filled quickly, whereas in traditional Aub-anbars, the water-entering cross-section is limited and much longer time is needed for the Aub-anbar to be filled.

5. Mud and dirt stay on the uppermost layer and clean water leaks down into the layer beneath the sands, whereas in traditional type of Aub-anbars, they would sink to the bottom and made a layer of sludge. The elements at the surface dry in a short time during summer and build an isolating layer against the heat of the sun. That layer can be removed in autumn and be used as fertilizer on farm lands and gardens. This needs less manual work than removing the sludge from the bottom of the traditional Aub-anbars. Moreover, the removed part can be replaced by sands to make the filtering layer more perfect.

6. To access the water in this type of Aub-anbars is only possible through piping and pumps. This in turn reduces the danger of pollution to its minimum.

7. Obviously, there is no natural water without any microbes. But, here two types of microbes are considered: aerobic microbes and anaerobic microbes. There is no doubt that in sandy Aub-anbars, the aerobic microbes are minimized, but to get rid of anaerobic microbes is impossible and they exist in water of all deep and semi-deep wells. Nevertheless, the water from those wells is consumable.

8. Chloride may be added to the water whether directly to the Aub-anbar or the pipes used for discharging the water. Other ways are also possible. We can adjust the grains of sand and pebbles and add chemical or mineral substances such as fluoride dicalcium and keep the water at its highest possible level of drinking [17]. Other methods to prevent the contamination of water and stop the entering of the floating particles and seeds of parasites into the Aub-anbar has been outlined below [31]:
9. The outer surface and the circumference of Aub-anbar should be made completely impenetrable up to 0.5 m into the ground, for collecting the rainwater.
10. Sand and pebbles are placed on the impenetrable layer where rainwater would enter. The impenetrable underneath layer has steep drainage canals where the rainwater after penetrating into the sand and pebbles gets into them.
11. The collected surface water, through canals, made after depositing pools; enter into sandy layers once again to lose its remaining suspended matters.
12. After three times of filtering, the rainwater enters into the Aub-anbar and it is stored there. To disinfect the water chemically, some chemical substances are added to the water.

In another method, instead of placing sand and pebble layers on the rain-collecting surface, they are dumped into Aub-anbar and let the rain flow through it. In this way, the whole structure acts like a small underground aqua-table [31].

4.2.2 Discharging and Dredging

Aub-anbars were cleaned while they were dredged. They were cleaned after discharging the remained water at the end of summer or hot season. A quick person was sent to the water-tap area and he skillfully dropped a special vertical piece of wood into the water which caused the water to flood into the water-tap area. Sometimes the water would rise to the midway of the staircase and put the life of the person in danger. In some Aub-anbars water-tap was kept open until the remaining water, usually muddy, to be evacuated [14].

In some Aub-anbars, under the tap, a hatchway called trunk, was made which would let the sunken mud to exit forcefully through and from the tap area to be drained into a well. In some cases, the hatchway was made in the center of the reservoir at the floor which was connected to a well. Opening the hatchway would let the muddy water to be drained into the well [5].

After the evacuation of the entire water, mud, sledge, and sedimentaries, water exit was tightly closed with a piece of willow wood covered with cotton and the fat of sheep to prevent any drainage. The expense for the cleaning of the reservoir was provided by the people in the neighborhood [32]. To clean the reservoir, some strong men were hired. Those men wore fluffy clothes while cleaning Aub-anbar. Since the use siphon for Aub-anbars, the cleaning task became easier [5].

Another time that Aub-anbars were cleaned was the late winter or during New Year time. In Mashhad, one of the traditional customs was the renewal of the old water in Aub-anbars as people renewed their clothes or their house furniture and the like.

The renewal of the water was carried out collectively with the help of all the families in the neighborhood. All people would fall in a line with their buckets in their hands. They would pass the filled buckets to each other while giving the empty one in return. This way the process of discharging the water was done quickly. The Aub-anbar was filled with fresh water over night. In villages, this was accompanied with playing music by local people, creating a happy atmosphere.

If Aub-anbars were not discharged and cleaned at proper times, they were gradually ruined. As it is mentioned in the writings of the past [3]: "people could not afford to clean Aub-anbars for a number of years. Therefore, the amount of sledge becomes so much during the years that the cleaning becomes something impossible in terms of expense. When the roof of Aub-anbar cracks and is not repaired and holes appear in it, the roof gradually falls down and Aub-anbar turns into a dirty ditch which may be used by trespassers and birds for a while and finally becomes a totally ruined and deserted place".

References

[1] Farasat Kish, A. (1995). Determining the Excrement Pollution of Aub-anbars in Kashan. Doctoral Thesis in Medicine, School of Medicine, University of Medical Science, Kashan, Iran.

[2] Ghafori, M. R., and Mortazavi, S. R. (1992). *Hydrology*. Tehran University, Tehran.

[3] Massarat, H. (2010). *Aub-anbars of Yazd*. Yazda, Tehran.

[4] Hasib al-Karaji, A. M. H. (1994). *The Excavation of the Hidden Waters*, translated by H. Khadivjam. The Research Center for Humanities and Cultural Studies, Tehran.

[5] Jahanfar, J. (1994). Water in old Iran. *Sokhan-e Daneshjo J*. 3 & 4.

[6] Jorjani, E. H. (1965). *Kharazm Shahi's Savings*, attempted by M. T. Danesh Pajouh and E. Afshar. Tehran University, Tehran.

[7] Erfanfar, J. (2007). An Anthology of Readings in the Texts of Donors in Iran. *J. Farhang Yazd* 30.

[8] Peernia, M. K. (1993). *An Introduction to Iranian Islamic Architecture*, ed G. H. Memarian. Iran University of Science and Technology, Tehran.

[9] Tabatabaee Ardekani, M. (2002). *Folk Dictionary of Ardakan*. Cultural Council of Yazd Province, Yazd.

[10] Besharat, H. (2007). *Yazd, My City*, 2nd Revision. Andishmandan-e Yazd, Yazd.

[11] Safa al-Saltaneh, M. A. (1987). *A Report of Kavir, Travel Account of Safa al-Saltaneh Naeinin*, attempted by M. Galin. Ettelaat, Tehran.

[12] Pour Ebrahim, H. (1992). *Gonabad Geography*.: Marandiz, Gonabad.

[13] Safeezadeh, M. (2004). *A Prism of Wonders,* Iran Newspaper: Tourism, 10th Year, No. 2863.

[14] Siroux, M. (1949). *Caravansérails d'Iran et petites constructions routiers*, MIFAO, Cairo.

[15] Farshad, M. (1997). *History of Engineering in Iran*.: Balkh, Tehran.

[16] Ghobadian, V. (2008). *Climatic Analysis of the Traditional Iranian Building*.: Tehran University Tehran.

[17] Farzad, H. (1970). Design of Sandy Aub-anbars. *Daneshmand J.*

[18] Nour Bakhsh, H. (2002). *Water Ponds in Persian Gulf, Farhang-e Mardam J*, 1, 78–91.

[19] Pour Moghaddas, H., Shah Mansouri, M. R., and Zafarzadeh, A. (1999). Chemical and microbial investigation of water in rural Aub-anbars in Golestan province. *Water Wastewater J. Isfahan* 45.

[20] Zafarzadeh, A. (1999). Chemical and Bacteriological Investigation of Water in Rural Aub-anbars Minoo Dasht City in Golestan Province. MS Thesis, School of Health Engineering, University of Medical Sciences, Isfahan, Iran.

[21] Mohammadi, Z., and Shah Mansouri, M. R. (2001). "Bacteriological Investigation of Water in Aub-anbars of Bandar-e Lengeh City, Hormozgan Province," in *4th National Conference on Environmental Health in Iran*, Yazd Shahid Sadoughi University, Yazd, Iran.

[22] Dehghani, A. R., Khani, M. R., and Mohammad Nia, M. (2009). "Study of the Quality and Quantity the Water of Cisterns of Province of South Khurasan-Birjand," in *12th National Congress of Environmental Health*, Shahid Beheshti University of Medical Sciences, Tehran, Iran.

[23] Karabi, H. (2008). Qualitative and Quantitative Investigation of Water in Sabzevar. BS Thesis, Department of Environmental Health Engineering, Islamic Azad University, Medical Sciences Branch, Tehran, Iran.

[24] Khani, M. R., Dehghani A. R., et al. (2007). "Study of Passive Cooling Systems and Chilling Systems Rule in Environmental Pollution Reduction," in *The 1st Conference and Exhibition of Environmental Engineering*, Tehran, Iran, 529.

[25] Khani, M. R., Yaghmaeian, K., and Dehghani, A. R. (2009). An experimental study in passive cooling systems and investigation of their role in diminishing energy usage and environmental pollutants. *Intl. J. Appl. Eng. Res. (IJAER)* 4, 519–528.

[26] Varjavand, P. (1987). *Iranian Architecture, Islamic Period*, attempted by M. Y. Kiani. Jihad-e Daneshgahi, Tehran.

[27] Jalali, A. (2009). Physical and Chemical Study of Water in Lar Aub-anbars. BS Thesis, Department of Environmental Health Engineering, Islamic Azad University, Medical Sciences Branch, Tehran, Iran.

[28] Kheirkhahe Arani, R. (2006). An Investigation of Indigenous and Cultural Buildings of Aub-anbars, Particularly in Kashan. *Farhang-e Isfahan J.* 31, 79–85.

[29] Farrokh Yar, H. (2007). *Aub-anbar, a Forgotten Memento*. Helm, Qum.

[30] Shahri Baf, J. (1992). *Old Tehran,* Vol. 3. Moein, Tehran.

[31] Monzavee, M. T. (2007). *Urban Water System*. Tehran University, Tehran.

[32] Hekmat Yaghmaei, A. (1991). *At the Shore of Salt Kavir.* Toos, Tehran.

5

Energy and its Storage

5.1 Introduction

Energy is one of the basic needs of our world. In fact, energy is the original driving force of human lives. The various stages of civilization of human beings have been formed based on the discoveries, inventions, and the use of different sources of energy. Therefore, the issue of energy may be viewed as the principles and pillars of the social life of man of today. Inventions, researches, and use of various kinds of energy have been the important steps which man has taken in the process of social advancements.

Physical strength had been the primary kind of energy in the very early days of human life on this planet. Later on, he could tame the animals and burn the wood to fulfill his needs. Not very long ago, he could access fossil fuels such as coal, oil, and natural gas and improve his technical and material life as never imagined before [1]. So, man's life has always depended on energy consumption. Changing the food into energy, he sustains his life and does some kind of work.

Energy has been defined as the ability to do some work. Work is specified as the force multiplied by the change of distance in line with the force. Energy exists in different forms, the most important of which are as follows [2]:

1. *Electrical energy* (electricity) is easily transmitted and has many applications like light;
2. *Mechanical energy* which is used in motor-vehicles and agricultural and industrial machines;
3. *Potential energy* like the water behind the dams for generating electricity, irrigation of lands, and supplying water for cities;
4. *Kinetic energy* like the energy of wind to generate electricity or mechanical energy for discharging water from the wells and rivers;

5. *Chemical energy* which is released in chemical reactions, like the energy produced by burning fossil fuels such as oil, natural gas, coal, wood, agricultural residues, waste, and the like;
6. *Nuclear energy* released in a nuclear reaction which is used to produce electricity in power plants;
7. *Internal energy* which is the total energy of an object at its molecular level;
8. *Heat energy* which is transferred from the object with higher heat to the object with lower heat.

The quality of energy varies from one kind to another, i.e. some kind of energies has a better quality like electrical energy, chemical energy, and nuclear energy, whereas ocean's heat energy and molecular energy have a lower quality [2]. Several factors like the lack of or emerging of new technologies cause changes in the use of energy. Therefore, it can be said that every energy system includes the producing technology and the way it is consumed. The purpose of each energy system is the distribution of energy to the consumer so that the consumer can make the most of it.

The services that an energy system offers are carried out by extraction or collection of energy in one or more stages. In fact, energy changes into a different product that is needed by the end-consumers. Of course, these services are the result of the combination of various technologies, investments, and technical knowledge in materials and primary energy sources (primary energy is the kind of energy which has not gone through any conversion process like crude oil). In brief, it is the energy which runs everything.

5.2 History of Energy

Although the concept of the term energy has been first noticed in works of Aristotle; nevertheless, because of many changes in books and rewriting of many of them, it is precisely clear that who has used this concept first. The experts and scholars in history, language, and etymology know that the word energy has been derived from the French word *energie,* which in turn was derived from Latin *energia.* This Latin word was itself derived from the Greek word *enorgus,* which is a combination of *en* and *orgus* meaning 'work within or inside' [3].

Energy is a fundamental physical quantity. According to modern theories of atomic structure, matter consists of minute particles known as atoms. These atoms represent enormous concentration of binding energy. According to the

'Theory of Relativity', energy in an object is equal to the amount of atoms of that object which is calculated through Einstein Equation ($E = mc^2$), i.e. if an object loses its energy (E) through radiation, its mass will decrease as E/c^2. Also, if an object loses a part of its mass, it releases that much energy which is equal to mc^2, where m is the lost mass.

The concept of energy is first observed as Vis Viva in the works of Leibniz and it was the result of mass of an object multiplied by its speed powered by 2. In Leibniz's idea, the total *Vis Viva* in the universe is constant. To solve the problem of speed-reduction because of friction, he defined 'heat' as the movement of the atoms in matter, in agreement with Newton. Despite all efforts, the concept of energy was not accepted until two centuries ago. It was in 1807 that Thomas Young used the term Energy instead of *Vis Viva*. In 1829, Gustave Gaspard Coriolis defined the Motion Energy. In 1856, William Thomson titled as Lord Kelvin called Motion Energy, and Kinetic Energy. In fact, there is no difference between these two terms, because the word kinetic is derived from a Greek word meaning *motion*. In 1853, the term Potential Energy as it is used nowadays was first observed in the works of William Rankine [3].

When steam engines emerged, engineers had to find formulas to define heat and mechanical equations of those systems. Julius Robert Mayer was the first who studied on the issue of heat and the conversion of mechanical energy to the other types of energy. He could specify the amount of energy produced by measuring a certain amount of mechanical energy. In fact, he measured something known as 'the equivalence of mechanical heat'. Moreover, Mayor could demonstrate that any type of energy can be converted into other types while keeping its total energy fixed. Mayer was soon followed by James Prescott Joule who after many experiments came to the conclusion that with certain amount of energy in hand, regardless of its type, we can always have the same amount of heat produced too. Furthermore, he found out that each type of energy can be converted into another type without any amount lost or gained. This shows that there exists the 'energy survival law'. Another scientist by the name of Hermann Von Helmholtz also found out that energy never perishes.

Mayer, Joule, and Helmholtz, who were the researchers of 1840s, proved that the energy never perishes and it always exists but may change to different types. In fact, the total energy in the universe will always be the same amount [3].

Nicolas Leonard Saadi Carnot was the first who noticed that the energy never perishes; nevertheless, all of it cannot be exploited for useful work.

He proved that a steam-engine cannot convert all the heat given to it, into mechanical energy. Some portion of the energy is wasted. Rudolf Julius Clausius continued the works of Carnot and did not limit his studies just to steam-engines. He studied on all kinds of works done by energy. He was the first one who gave a precise definition of the term 'work'.

Regarding what went before, it seems necessary to discuss the 'first and second laws of thermodynamics' in brief. The first law is called the 'energy survival law' [4]. According to this law, in each cycle (any change of initial state of a system and back to its original state) that the system (a specific mass of matter which can be identified from the other types of matter) goes through, 'heat cycle integral' is equal to 'work cycle integral'. There is no limit to the flow of heat and work, considering this law. The first law is equally applicable to any cycle of energy converted into work and vice versa. Nevertheless, practically and by experience, it is understood that if any cycle, not deterring the first law, really functions. The experimental evidence had led the scientists to formulate the second law in thermodynamics. The second law confirms that the processes (passing of a system through the consecutive states) go in a certain direction and not against it. The second law brings about a feature (a quantity which depends on the system state with no relation to its path to gain that feature) called 'Entropy' which makes it possible to apply the second law, quantitatively in processes.

5.3 Energy Sources

Energy sources can be divided into two groups:

1. Renewable energy;
2. Non-renewable energy.

5.3.1 Renewable Energies

Renewable energy sources are the ones that man has no role in their process of production, like solar energy, wind, and others whose origin is outside our planet; or the earth heat energy, the heat which is originated by the inner most part of the earth. These sources, compared to the man's civilization history on the earth are almost non-perishing. Nevertheless, one day, the sun which is the origin of these types of energies will die. The renewable energy sources are the ones which can reduce the problems of contaminants, the increase in earth temperature and the density of carbon dioxide in the atmosphere. That is why, their role in supplying the energy needed on this planet is gradually increasing.

The primary cost of systems using renewable energy in one hand, and the cheap oil and natural gas on the other, had been obstacles in using and developing the renewable energy sources. But, the rise of oil prices in 1973 made the industrial countries to pay more attention to the production of energy from non-fossil fuels.

The renewable energy sources are almost non-ending, easily accessible, and do not damage the environment. These advantages have encouraged many countries to try to use them in one way or another depending upon their climate condition and the technology available. Solar, wind, biomass, earth, heat, waves, tides, heat of seas and oceans and hydraulic are among the renewable and clean type of energies, which are mostly attended by at present.

Not only is the sun a great source of energy; but also is the origin of life and all other types of energy. Based on the scientific estimations, about 6,000 million years has passed the birth of the sun and 4.2 million tons of its mass turns into energy per second. The most important gases of the sun include Hydrogen (86.8%), Helium (3%), and 63 other elements, the most important of which are Oxygen, Carbon, Neon, and Nitrogen [5]. The diameter of the sun is 1.39×10^6 km. Regarding the tremendous weight of the sun which is about 333,000 times earth weight, can provide solar energy for 5 billion more years. The temperature of the sun at its inner most part is approximately 10 to 14 million °C and at the surface is approximately 5600 °C. The surface energy of the sun emits into space as electromagnetic waves.

Our planet Earth is 150 million km away from the Sun. It takes 8 min. and 18 s for the light from the sun to reach the earth. The earth's share of energy from the sun is approximately $1/2 \times 10^9$ out of its total energy.

The use of solar energy for different purposes goes back to prehistory period, probably to the era of pottery. In the old days, the clergy in temples, using large polished bowls or plates and focusing the sun light, could ignite fire-place of the altars. One of the Faros of Egypt could build a door which opened and closed by the sun shine and the sun set. The most interesting story about the use of the sun, relates to Archimedes, the great scientist and inventor of the old Greece. It is said that he could destroy the fleet of the Romans by the heat energy of the sun. He set a number of square-shaped mirrors on a movable stand and focused the heat of the sun on the ships and set them on fire. The Iranian traditional architecture paid particular attention to the appropriate and effective use of the sun's energy [5].

At present, the energy of the sun is utilized in many different ways. For example, photo-electric cells which turn the light of the sun into electricity. These cells which are known as solar-cells can be used in simple systems such as light-energized calculators, wrist watches, and in more complicated systems like house lights, communication equipment, and in satellites.

The wind energy like other types of renewable energies can be used in many places in a scattered or de-centralized ways in different parts of the world and it is almost always available, though this natural energy is not always constant and can vary depending on the time of the day or even in different seasons. Man has been using this type of energy for thousands of years for his mills.

Wind energy, before the Industrial Revolution, was widely used [6], all over the world; but, after Industrial Revolution, utilization of fossil energy became common place because it was cheap and reliable and substituted wind energy. The old wind turbines were economically no match for the turbines which are run by oil and natural gas. Nevertheless, two great oil shocks in 1973 and 1978 cut down the cost of energy by wind turbines compared with global energy cost. It was then that research centers throughout the world tried to investigate different technologies to exploit wind energy. In recent years, the environmental problems and the issue of green gas and extreme alteration in climate due to the use of fossil energy have intensified the need for wind energy.

Man has discovered the ceaseless energy of the wind from thousand years ago and has taken advantage of wind energy to run ships and wind mills. It was known through the years that wind energy can be converted into mechanical or electrical energy. Historical evidence demonstrates that the construction of wind mills in Iran, Egypt, Iraq, and China goes back to very many years ago. In these countries, wind mills were used to crush grains and pump water. In 1700 BC, as it is understood from historical proofs, King of Babel, Hammurabi, proposed a plan according to which the fertile plane of Mesopotamia could be irrigated by using wind energy [6]. The wind mills built in those times had a vertical axis and are similar to the ones found in Khaf, Nahbandan, and Sistan in Iran (Figure 5.1).

According to majority of historians and researchers, the technology of wind mills originated from the eastern and south-eastern parts of Iran, i.e., Khorasan and Sistan [7–9]. In those regions, there is always an ever-lasting wind called 120-day wind. The earliest written document, in which there is a mention of wind mills, is an old Indian book, *Arthosastra of Kantilya*, from 1400 BC. In this book, there is a mention of water ascending [8, 9]. Iranians were the first who used vertical wind mills to crush the grains in 200 BC.

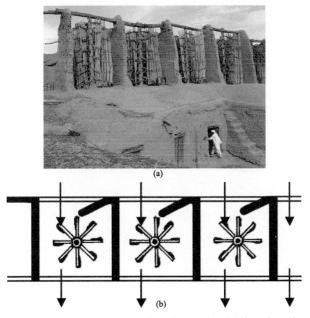

Figure 5.1 (a) a view of a series of wind mills with vertical blades; (b) a sketch of top-view of vertical wind mills [12, 13], in Nashtifan, Khaf, Khorasan Razavi Province, Iran.

Regarding the land of wind mills, Massoudi (947 AD) says that [10]: "the land of Sistan is the land of wind and sand and it is the city that wind runs the mills, pulls up the water from the wells, waters the gardens and there is no city in the world that could have been benefited from the wind as Sistan". The well-known Iranian geography scientist, Estakhri, (about 919 AD) confirms Massoudi and states that [9]: "there is a strong wind, and wheels have been made that turn by that wind".

The people of old cities of Iran such as Yazd, Kashan, and Kerman used Baudgeers to direct the wind into the houses (Figure 5.2). Bahadori and Dehghani talk about Baudgeers [11]: "Baudgeers are the structures which had been used since centuries ago in some regions of the Middle East and Egypt where it was arid and dry or wet. Baudgeers were used to ventilate the houses. In fact, the role of Baudgeers was to direct the outside air into the buildings and passively cool the houses or the work places. These structures were also used in public water reservoirs (Aub-anbars) to keep the water cool in hot seasons of the year". They believe Baudgeers are originally from Iran and from there they had been copied in other countries of the Middle East and Egypt.

Figure 5.2   A view of some Baudgeers in old part of Yazd city, Iran.

At present, wind energy is exploited in different ways in different parts of the world. For example, wind turbines are one way of using the energy of the wind. Since 1975, there has been a great advance in the use of wind turbines to produce electricity. It was in 1980 that the first wind-electricity turbine was connected to the main lines of electricity in the USA. A short while after, the first farm, run by the electricity produced by wind, with a few mega watts, came into operation in the USA too.

By the end of 1990, the capacity of the wind-electrical turbines connected to the main lines of electricity reached at 200 MWh, which could produce 3200 GWh electricity in a year. Almost all this amount of production relates to California in the USA and Denmark. Nowadays, other countries like Netherland, Germany, England, Italy, and India have huge projects in hand to produce wind energy at national scale and even for export [6].

5.3.2 Non-renewable Energies

The sources of non-renewable energies are the sources which through various processes by man convert into heat. The heat produced in a heat engine changes into work. Among these sources are wood, fossil fuels, and nuclear energy which through combustion or breaking of atoms produce heat. Burning of the wood to warm the house and breaking of the atoms in a reactor of a modern power plant are some of these processes. One big disadvantage of these energy sources is their damaging effects on the environment while the process is being carried out and their limitations as well. Of course, meanwhile, the search for finding new sources and employing less polluting processes is continued. The identified oil and natural gas reserves worldwide and the rate of consumption, seems to be enough for the next 40 years. Although it is possible to provide liquid fuels from coal and procure energy for longer time, its environmental impacts are catastrophic [2].

The non-renewable energy sources can be divided into four groups [14]:

1. Solid non-renewable energies such as wood and coal;
2. Liquid non-renewable energies such as oil and benzene;
3. Gas non-renewable energies such as natural gas;
4. Nuclear energy.

Figures 5.3–5.5 indicates the production rate of main non-renewable energies.

Oil, coal, and natural gas are among the conventional energy sources in the market. Fossil fuels are considered the principal source of energy in the world. Transportation vehicles, power-plants, industries, and commercial and house heating systems emit a huge amount of dangerous gases and contaminants in the air which cause various diseases and endanger the environment.

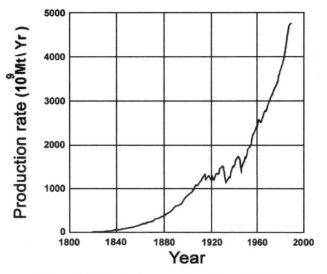

Figure 5.3 Production rate of coal in world [15].

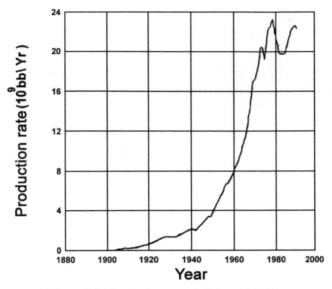

Figure 5.4   Production rate of oil in world [15].

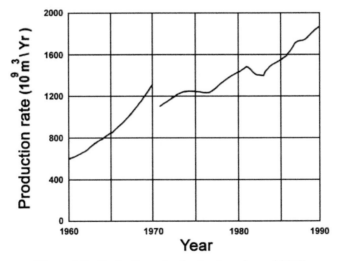

Figure 5.5   Production rate of natural gas in world [15].

5.4 Energy and Environment

Today, the provision of energy is accounted as the basic requirement for the economical and social development of countries. Population changes and the growth of city life in one hand, and the deficiencies in the process of production, transfer, distribution, and consumption on the other, and lack of enough depending on the secure and clean energies, all have caused a rising demand for energy and speedy exploitation of the resources available. While the ways of procuring and producing energy is a significant factor in polluting the environment, the quick consumption of non-renewable energy resources and the increase in various kinds of pollutions have brought about the crises of energy and environment in the third millennium [16].

The processes of the production and the use of different types of energies are determining factors in the environment pollution at the local, national, and international scales. Considering all these, focusing on the emission rate of polluting gases and green gas phenomenon and the investigation of their changing trend at different periods are appropriate means for planning and making policies to reduce the negative effects of energy consumption.

5.4.1 Air Pollutants

In the air surrounding us there are substances which could harm the plants, animals, and human beings. These elements which are known as airborne contaminants are generated both by natural processes and human activities. The concept of air pollution is vast and depends on chemical, physical, or biological factors which change the natural characteristics of the atmosphere. The atmosphere of the Earth is a complicated and active phenomenon, which causes the birth of life on the Earth. Any change in this natural phenomenon can endanger the life on this planet. Since some decades ago, for instance, man has noticed the reduction in stratosphere ozone, a consequence of the air pollution which has damaging effects on the life on the planet and imposing a threat on the human lives.

There are many reasons for the pollution of the air. One of the most important of them is the use of fossil fuels, for instance, the use of them in power plants to generate electrical energy or in cars to make them run. It is estimated that 70% of the air pollution in Tehran is caused by the gases emitted from the cars [2].

The polluting sources of the air may be divided into two groups:

1. Natural sources the most important of which include:
    1.1 Storms and dust;
    1.2 Volcanic activities;
    1.3 The smoke and ashes caused by forest fires;
    1.4 Meteorites;

    1.5 Plant and animal sources;

    1.6 Warm mineral springs.

2. Man-made sources the most important of which include:

    2.1 Motor vehicles;

    2.2 Industries and power plants;

    2.3 Commercial and house heating systems;

    2.4 Waste-material burners;

    2.5 Radioactive matters.

Air pollutants are also divided into the first and the second types. The first type is the pollutants which are released into the air through the man-made or natural pollutants. From this type, carbon mono dioxide and carbon dioxide can be mentioned. The second type is the ones produced by the chemical reaction of the first-type pollutants with the other elements of the natural air. The creation of ozone and photo-chemical smoke are the significant examples of this type.

The most important pollutants of the air include:

1. Sulfur combinations;
2. Particulate matter (PM);
3. Nitrogen oxides;
4. Carbon monoxide;
5. Ozone;
6. Gaseous organic combinations;
7. Hydro-carbons.

5.4.2 Ways to Fight Air Pollution and Environment Pollution

The most important ways proposed to prevent or reduce the climate change and destruction of the environment include:

5.4.2.1 Economizing energy

Inappropriate use is one of the factors in the increase in energy consumption in developing countries. Statistics show that, in Iran, the use of energy has risen from 392 million barrels of oil in 1986 to 1356 million barrels in 2006 (6.4% annual growth) [16]. In order to save the energy, house appliances should change into systems with high efficiency and the old technology in industries be substituted or renovated.

5.4.2.2 Use of alternatives

Since the most noticeable damage caused by using fossil fuels is the concentration of carbon dioxide in the atmosphere, the first step in substituting them is the use of fossil fuels with low carbon instead of fuels with high carbon, like natural gas instead of coal, benzene, and oil. High-carbon fuels emit carbon dioxide and sulfur dioxide, which play a significant role in the pollution of the air, whereas, in the consumption of natural gas, their amount diminishes to a great extent. However, it should be noted that if it is used widely, although carbon dioxide decreases considerably, nevertheless, the amount of methane and nitrogen dioxide increases.

At present, many countries, particularly the ones not possessing fossil fuels, take advantage of nuclear power; it provides 17% of the global energy supply. There are advantages to the use of nuclear energy, but there are dangers as well. The incidents of Chernobyl nuclear power plant in former Russia, Three Mile Island in the USA, and Fuchu Shi Ma in Japan have been catastrophic and many lost their lives and many more were impacted with radioactive substance. Moreover, vast land areas have been contaminated with the radioactive substances; and still there is the problem of nuclear waste that keeps active for years and should be safeguarded in special storage places.

5.4.2.3 Enhancement of energy efficiency

In the beginning of the 20th century, the efficiency of electricity production was 5%. At present, the average global conversion of primary energy to secondary energy (the kind of energy which is obtained through the conversion of primary energy like electricity) is around 74%; however, the average conversion of secondary energy (the kind of energy practically demanded by consumer for heating, lighting, and motors) to useful energy, i.e., the consumers' efficiency, does not exceed 46%. Therefore, the total efficiency conversion to useful energy is only 34%. Obviously, the enhancement of efficiency would reduce the fuel consumption and the emission of pollutants.

5.4.2.4 Exploitation of renewable energy sources

As it was mentioned before, the most important advantages of using renewable energies relate to the facts that they are virtually unending; they are easily accessible; and their exploitation does not damage the environment. Regarding those points, many countries today, considering their climatic condition and the advancement of technologies in using clean and renewable energies, have made a lot of efforts to exploit them. Solar energy, wind, biomass, earth's heat, waves, tides, and hydraulic fuels are among the clean and renewable sources of energy.

5.5 Energy and Development

With the development in economical sector, the consumption of energy increases visa vie and it is expected that its consumption in next decades would increase accordingly. Even the optimization and advancement of available technologies would not stop the rise in the consumption of energy. However, efforts are being made to reduce the use of energy, which might alter the trend of present energy consumption in future [2].

Energy is an important input in all sectors of any country's development in terms of economy, social, environmental, and human resources. The standard of living of a given country can be directly related to per capita energy consumption. In order to have a sustainable development in the global economy and have more prosperous nations, we should become aware that the kind of energy used today and their ways of exploitation must change in future; otherwise, the unmethodical use of energy would impose extreme damages to the environment, inequality would increase, and the global economic growth would suffer [2, 17].

Nowadays, the use of energy has brought about so much environmental problems that the prosperity of human on this planet has been negatively influenced. The crises in political, economical and similar problems, the limitation of fossil fuels, the growth of population, and too much consumption all have become global issues and they extend more and more every day. Clearly, the economic and political support of countries depends upon their use of fossil fuels and the cessation of those resources is not only a threat for the exporting countries; but also, a serious difficulty for the importing countries of fossil fuels. The new conditions, has made the sustainable development a serious issue among strategic planners in all countries in different sectors.

To produce energy and find reasonable ways of its exploitation to develop communities in long run, in all social and economical dimensions and paying attention to environmental issues, is what is known as sustainable energy. This does not mean that the access to sustainable energy is easily feasible; rather, it means that the aim of energy production and its use should be in such a manner that the development should be parallel to the prosperity of man in the long run, while having a balanced environment and a sustainable improvement.

The term sustainability is extensively applied to explain a world that the human and natural systems to be able to continue their existence up to a far future. The concept of development, as referred to in Chapter 1, means to propose solutions for the common styles of social and economical activities so that to prevent the destruction of natural resources and sources, the demolishing of ecosystems, pollution, the irregular growth of population, injustice, and the diminishing of the life quality. The realization of sustainable development in every country depends on the domestic capabilities, human resources, and technology available and financial resources in that country.

At present, if the existing technologies stagnate and consumption grows, the energy use would not be sustainable. In 1992, in the general assembly of UN, in a statement, it was announced that there is a relationship between energy and the global warming and, in 1997, energy and transport were recognized as two important indexes in a sustainable development. Moreover, in the same year, the Kyoto Protocol, regarding the preservation of environment due to energy consumption was reached. Different industries also, have deeply understood the importance of optimization in the use of energy. In 1998, The Global Energy Association discussed the different ways of energy procurement and the impact of energy consumption on the environment at various levels [2]. There are not any physical limitations, at least in next 50 years; however, the present system of energy provision faces many challenges in terms of environment, economy, and political disputes at regional and international scales [2, 18].

The lack of a sustainable energy system at present may be attributed to the following reasons [2]:

1. The new fuels and electrical energy are not proportionally available for all and this inequality causes disputes in ethical, social, and political dimensions in the world and the situation gets steadily worse.
2. The present system is not much reliable or man is still unable to develop it in all different ways. One-third of world population is not capable of producing enough energy, and still 1/3 suffers from the lack of sustainability of energy.
3. The negative impact of energy consumption at local, regional, and global levels has become noticeably significantly enough in putting the lives and prosperity of communities in danger.

Iran, a country rich in oil, while having 1% of the world population, consumes 9% of oil products in the globe [19]. In Iran, the rate of growth of energy consumption within a 16-year period had been on the average 5.8% annually. Compared to global rate, it is seen that the average energy consumption in the world within a 10-year period (1984–1994) was around 15%. In recent years, the energy consumption in Iran has been 5 times that of world average [2].

The estimations in 2006 shows that the percentage of electrical energy consumption in Iran, compared to total consumption in the world, had been 8.6% in that year. The rate of growth of power consumption is 8.2% annually in Iran. The increasing growth of energy consumption in residential, public, business, agricultural, and transportation sectors in 2006 had been 10.9, 10, 3, 7.9, and 6.4%, respectively, compared to 2005 [16].

Available statistics and information indicates that the production of primary energy in 2006 was equal to 2327.8 barrels of crude oil, of which 68.5% was crude oil, liquids, liquid gases, and added materials, 29.6% natural gas, 1.1% solid biomass, 0.5% water and renewable energies, and 0.3% coal. The energy had been provided by oil products (48.4%), natural gas (40.1%), power plants (8.6%), solid biomass (2.5%), and coal (0.3%) [16].

In 2006 in Iran, the percentage and the pollutants originated from all different energy-consuming parts were: the transportation with solid particle material (77.9%), NO_x(63.4%), CH (96.7%), and CO (98.9%); Petrol with carbon monoxide (97.8%); Gasoil with flowing particles (75.1%); natural gas combustion with carbon dioxide (47.7%). Gasoil is the main source for the 60.4% of SO_3 and 61.5% of SO_2 produced in the country. The per capita emission of Carbon dioxide is 6 tons annually [16].

The available data related to the emission of pollutant and green gases from various consumed fuels in 2006 are shown in Table 5.1 and the share of each type of fossil fuels in emitting those gases in Table 5.2.

Table 5.1 reflects the fact that the maximum emission of NO_x, SO_2, SO_3, and SPM relates to Gasoil; CO_2 relates to natural gas; and CH relates to benzene. The maximum amount of gas among green gases relates to CO_2 in 2006.

The important point worth mentioning in Table 5.2 is that compared with other fossil fuels, natural gas is considered a clean fuel; and diminishing of floating particles, because of the high use of this gas, among the other fossil fuels, is noticeable.

Table 5.3 reflects the amount of the emission of the pollutant and green gases out of the total energy consumed in 1967–2006 in tons in Iran. As the table shows, the variation in carbon monoxide emission is more considerable than the other pollutants during the same years. In that period, carbon monoxide, carbon dioxide, and floating particles had become 32.5, 26, and 17 times more respectively.

Table 5.4 indicates the emission of the pollutant and green gases out of the total energy consumption in kilogram per head during 1967–2006. According to the table, per head emission of NO_x had increased from 2.4 kg in 1967 to

Table 5.1 Emission rate of pollutant and green gases from the conventional fuels in 2006 in tons [16]

Fuel/Gas	CH	CO	SO_3	CO_2	SO_2	NO_x	SPM
Fuel oil	6264	58	3743	46637228	245018	118673	15662
Gas oil	419989	137241	6060	83223992	514916	575166	263774
Kerosene	–	5610	–	17371096	17263	3596	–
Petrol	1692621	9403450	–	62412043	40301	362705	34926
Gas Liquid	1117	16471	–	7435154	36	2046	–
Natural Gas	5550	13700	–	200675413	670	251849	21347
ATK[1]	25423	8320	231	3060029	19414	31201	15254
JP4[2]	6231	34615	–	229745	148	1335	129
Total	**2157196**	**9619466**	**10034**	**421044699**	**837767**	**1346571**	**351091**

[1]ATK: Aviation Turbine Kerosene.
[2]Jet Propulsion Fuel.

Table 5.2 Fossil fuels' share in the emission of pollutants and green gases in 2006 in percent [16]

Fuel/Gas	CH	CO	SO_3	CO_2	SO_2	NO_x	SPM
Fuel oil	0.3	Negligible	37.3	11.1	29.2	8.8	4.5
Gas oil	19.5	1.4	60.4	19.8	61.5	42.7	75.1
Kerosene	–	0.1	–	4.1	2.1	0.3	–
Petrol	78.5	97.8	–	14.8	4.8	26.9	9.9
Gas Liquid	0.1	0.2	–	1.8	Negligible	0.2	–
Natural Gas	0.3	0.1	–	47.7	0.1	18.7	6.1
ATK	1.2	0.1	2.3	0.7	2.3	2.3	4.3
JP4	0.3	0.4	–	0.1	Negligible	0.1	Negligible
Total	**100.0**	**100.0**	**100.0**	**100.0**	**100.0**	**100.0**	**100.0**

19.1 kg in 2006, similarly, SO_2 from 4.1 to 11.9, CO_2 from 607 to 5972.6, and CO from 11.2 to 138.5 in those 39 years of time.

5.6 Energy and Economy

Energy plays a significant role in economy and production. The important issues relating to energy include [16]: trend of energy prices, the major index of energy consumption (consumption per head, the intensity of energy, the coefficient of energy, and the efficiency of energy), energy subsidies, and the inflation effects of energy prices.

The estimations in 2006 show that the subsidy for the energy in country was 383,162 billion Rials and the per head subsidy for energy was approximately 5435.2 Rials[1]. This shows 7.2% increase compared with 2005. In 2006, the

---

[1]One dollar was equal to approximately 10,000 Rials in 2006.

Table 5.3   Emission rate of pollutant and green gases out of the total energy of Iran between 1967–2006 in tons [16]

Gas/Year	CH	CO	SO_3	CO_2	SO_2	NO_x	SPM
1967	80395	296064	1442	16079158	108756	63994	20399
1971	129179	485940	2242	24772304	168474	99733	31521
1976	334831	1344073	4586	47904550	348844	231609	71204
1981	399661	1527543	6411	31894388	482279	306754	95103
1986	621684	2328184	10603	93182154	784845	489036	154068
1991	807221	3092375	11974	170110277	850443	629904	192243
1996	1065138	4263156	15303	240353433	1144295	814698	236419
2001	1417623	5989137	15653	302315645	1174945	994424	272013
2004	1822037	7956643	8413	356097025	713143	1168386	313426
2005	1988705	8749132	9113	381937529	768793	1256222	335148
2006	2157196	9619466	10034	421044699	837767	1346571	351091

Table 5.4   Emission of the pollutant and green gases out of the total energy of Iran per head between 1967–2006 in kilogram per head [16]

Gas/Year	CH	CO	SO_3	CO_2	SO_2	NO_x	SPM
1967	3.0	11.2	0.05	607	4.1	2.4	0.7
1971	4.3	16.5	0.07	840	5.7	3.8	1.0
1976	9.3	39.8	0.1	1421	10.3	6.8	2.0
1981	9.7	37.4	0.1	718	11.8	7.5	2.0
1986	12.6	47.0	0.2	1884	15.8	9.9	3.1
1991	14.4	55.0	0.2	3046	15.2	11.3	3.4
1996	17.7	71.0	0.2	4002	19.0	13.5	3.9
2001	21.9	92.8	0.2	4685	18.2	15.4	4.2
2004	27.0	117.9	0.1	5277.3	10.6	17.3	4.6
2005	29.0	127.8	0.1	5578.4	11.2	18.3	4.9
2006	30.6	138.5	0.1	5972.6	11.9	19.1	5.0

maximum subsidies of energy went to Gasoil with 28.2% and the least to liquid-gas with 2.9% [16].

By using passive cooling systems like Aub-anbars, it is possible to reduce the energy consumption and environment pollutants and devote the subsidies of those to other areas in need.

5.7 Methods of Energy Storage

There are several methods to store the energy at its lowest need and using the stored energy at the highest need. These methods include [20, 21]:

1. Mechanical energy storage

　　　　1.1　Pumped hydroelectric storage
　　　　1.2　Compressed air
　　　　1.3　Flywheel

　　2.　Electrical storage: the lead acid battery
　　3.　Chemical energy storage

　　　　3.1　Hydrogen
　　　　3.2　Ammonia
　　　　3.2　Reversible chemical reactions

　　4.　Electromagnetic energy storage
　　5.　Thermal (heat) energy storage

　　　　5.1　Sensible heat
　　　　5.2　Latent heat
　　　　5.3　Chemical reactions

　　6.　Biological storage.

Since the book tackles the issue of Aub-anbars, only the storage of thermal energy is discussed here.

5.7.1 Thermal Energy Storage

Thermal energy storage may occur through the increase or decrease in the temperature (change of sensible heat) or the change of phase of (change of latent heat) a substance or a combination of both. The storage of thermal energy is a temporary storage of energy at high or low temperatures for future use. Examples of thermal energy storage are, absorbing solar heat energy by collectors during the day and use of it at night; storage of the snow and ice in the winter to be used in summer; and storage of cold energy by chillers off peak and use of it at peak hours.

The thermal energy storage in Aub-anbars is an important means to satisfy a part of energy needed. This way we can reduce the use of fossil energy.

A complete cycle of energy storage includes: charging, storage, and discharging process. Storage of thermal energy could occur in short-term (within 24 hours) or in long-term (within a year) period. Chillers and Aub-anbars are examples of short- and long-term storage, respectively. The size of the storing element of energy depends on the cycle of the energy storage, i.e. short or long periods.

There are different storage media such as air, oil, etc. Nevertheless, it seems that water is one of the best media for the storage of sensible energy because of its abundance, low cost, availability, hygienic condition, high specific heat, and high volumetric heat capacity compared with the other liquids in the environment temperature ($\rho C_{Pw} = 4044\,\mathrm{kJ/m^3\,K}$). Since the storage of energy is in fact a delay in the time of energy storage and its use in future, water is an excellent environment to transfer energy from the stored location to the site of use.

5.7.2 The Criteria to Assess Energy Storage Systems

Among the criteria to assess the energy storage systems, the size and the cost are two important ones. Short- and long-term storage periods are other important factors that should be taken into consideration when choosing the storage system for different purposes. Aub-anbars being applicable in hot and arid regions as well as other places possess good characteristics, which make them quite reasonable in terms of cost versus energy use. Besides, they can store energy in huge quantity and for a long time.

References

[1] SANA. (2007). *What Do You Know About New Energies? Biomass Energy*, 4th Report. Iran's Organization of New Energy (SANA), Ministry of Energy, Tehran.

[2] Bahadori, M. N., and Yaghobi, M. (2007). *Natural Cooling and Air-Conditioning in Traditional Buildings in Iran.* Nashr-e Danshgahi, Tehran.

[3] Asimov, I. (1978). *Energy*, Translated by E. Saadat. Fatemi, Tehran 2nd Impression.

[4] Van Waylen, G., Sontag, R., and Bourgenak, K. (2004). *Basics of Classic Thermodynamics,* Translated by H. Haghighi Tajvar. Center of Salekan Cultural Services, Tehran.

[5] SANA. (2007). *What Do You Know About New Energies? Solar Energy*, 1st Report. Iran's Organization of New Energy (SANA), Ministry of Energy.

[6] *What Do You Know About New Energies? Wind Energy*, 3rd report. Iran's Organization of New Energy (SANA), Ministry of Energy, Tehran.

[7] Forbes, R. J. (1955–1972). *Studies in Ancient Technology*, Vol. 9, Leiden.

[8] Singer, C. A., Holmyard, E. J., and Hall, R. A. (eds). (1954–1965). *A History of Technology*, 5 Vols. England; Oxford.

[9] Farshad, M. (1984). *A History of Iranian Engineering*. Gouyesh, Tehran.

[10] Masoudi, A. A. (1958). *Moravvej al-Zahab va Meaden al-Johar*, Vol. 1 & 2, Translated by A. Payandeh. Ketab-e Tehran, Tehran.

[11] Bahadori, M. N., and Dehghani, A. R. (2008). *Baudgeer, A Masterpiece of Iranian Engineering*. Yazda, Tehran Iran.

[12] Dehghani, A. R. (2009). *Water in the Plateau of Iran: Qanat, Aub-anbar, and Ice-maker.* Yazda, Tehran.

[13] Moghtader, M. R. (1982). *Ab-anbar: Conservation de l'eau Sur le Plateau Iranian*, UNESCO, Paris, December.

[14] Rai, G. D. (2006). *Energy Sources*, 4th edn. Khanna Publishers, India.

[15] Sukhatme, S. P. (2000). *Solar Energy*. Tata Mc Graw Hill, New Delhi.

[16] *Energy Balance-Sheet in 2006*, Planning Office of Power & Energy, Ministry of Energy, Deputy of Power and Energy Affairs, 2006.

[17] Wohlgemuth, N., and Missfehlt, F. (2000).The Kyoto Mechanisms and the Prospects for Renewable Energy Technologies. *Solar Energy* 69, 305–314.

[18] World Energy Council. (1994). *New Renewable Energy Resources, a Guide to the Future*. Clays Ltd.

[19] Askari, M. (2002). *The Analysis of Energy Role in Major Economy of Iran between 1972–2000,* 4th Conference of National Energy, Tehran.

[20] Dehghan, A. A., and Dehghani, A. R. (2003). *An Experimental Evaluation of Thermal Stratification in Cisterns,* A Research Project, School of Mechanical Engineering, Yazd University, Yazd, Iran.

[21] Dehghani, A. R. (2007). *Analytical and Numerical Investigation of Heat Transfer in Cisterns,* MS Thesis, Department of Mechanical Engineering, Islamic Azad University, Science and Research Branch, Tehran, Iran.

6

An Experimental Study of Heat Transfer in Aub-Anbars

This chapter is devoted to the experimental investigation of the thermal performance in Aub-anbars. For this reason, an Aub-anbar was chosen, in a hot and arid central region of Iran located in the city of Yazd. The understudied Aub-anbar is known as Koocheh Biouk Aub-anbar, built in 1835. The diameter of the storage tank, which is equal to its height with a cylindrical shape, is equal to 12 m. The cistern has four wind towers at its four corners, each with the height and cross-sectional area of 10 m and 1 m^2, respectively. The ceiling of Aub-anbar is dome-shaped with an aperture on the top (Figure 6.1). The water comes out of a tap located about 90 cm above the reservoir bottom in order to prevent sediments created during the past years blocking the water way out of the tap. This Aub-anbar was donated by the Muslims in the area and has two separate staircases, one for Muslims and the other one for Zoroastrians (Figure 6.2).

6.1 The Explanation of the Experiment

In a research project, Dehghan and Dehghani experimentally measured the temperature of the cylindrical tank of Koocheh Biouk Aub-anbar in vertical and radial directions [1]. In winter (Jan. 2002), they filled the Aub-anbar from a nearby well with 15°C water. It should be mentioned that 2/5 of the Aub-anbar was already filled with water.

The water feeding the Aub-anbar was through an aperture on the top of the reservoir; therefore, the temperature distribution, in the beginning and after filling up the Aub-anbar, had been uniform with a thorough mixing of the water. The storage system was left intact until mid-April. Then, ten days after, the temperature distribution of the tank was measured every 10 days at 8 am and 8 pm. To measure the temperature distribution in vertical direction, 36-temperature sensors, each 0.3 m apart, were placed inside the tank. Also,

Figure 6.1   A picture of the Aub-anbar under study with its Baudgeer.

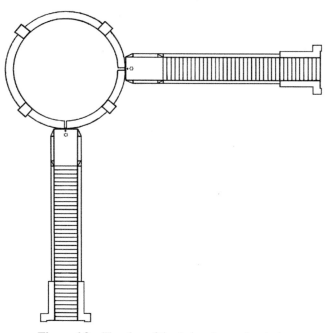

Figure 6.2   The plan of the Aub-anbar under study.

in the last two months of water drain (mid-August and mid-October), 3 thermometers were placed in radial direction with 0.5 m distance above the water surface to measure the temperature variations of that layer. Two twenty-channel digital temperature displays were used to collect data. Outside ambient temperature data were also obtained from Yazd Weather Bureau.

In order to simulate the water taking of the residents, the water discharge was done through a brass-made tap and according to the time table and the pattern presented in Table 6.1. The assumption was also that the storage tank could have supported 1000 residents in the local community with the daily average water consumption, in different months of the year, based on the same table. The discharge time of water, according to the discharge pattern in Table 6.1, was calculated through the following relations [2]:

$$\Delta t = \left[h_1^{\frac{1}{2}} - h_2^{\frac{1}{2}} \right] \times \left[\frac{2D_t^2}{d_i^2} \right] \times \left[\frac{1}{\alpha} + 8f \frac{\left(\frac{\sum L_e}{d_i} \right)}{2g} \right]^{\frac{1}{2}} \quad (6.1)$$

where h_1, height of water in primary state (m); h_2, height of water in secondary state (m); D_t, diameter of tank (m); d_i, diameter of pipe (in outflow of tank) about 0.0127 m; f, coefficient of friction (0.493); α, coefficient of correction (about 1 in SI system); and $\sum L_e$, equivalent length of outlet pipe (around 25 cm).

6.2 Results of the Experiment

As mentioned before, to simulate the real water discharge from the Aub-anbar by the residents, the process was carried out based on Table 6.1. The discharge of the chilled water from the tank through the brass-made tap was done in

Table 6.1 Average water consumption for days and months [1]

Month	Daily Discharge per Head (l/day)	Monthly Discharge per Head (l/month)	Monthly Discharge for 1000 People
Mid-April to mid-May*	4	124	124,000
Mid-May to mid-June	5	155	155,000
Mid-June to mid-July	6	186	186,000
Mid-July to mid-August	6	186	186,000
Mid-August to September	6	186	186,000
Mid-September to mid-October	4	120	120,000

*No discharge of water in mid-April to mid-May.

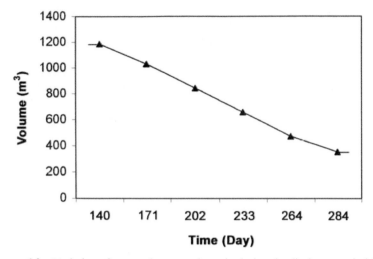

Figure 6.3   Variation of reserved water volume in during the discharge period [1].

mid-May through mid-October which is the period for the discharge, shown in the table. Figure 6.3 shows the reduction of the reserved water during the discharge time (consumption period). Also, Figure 6.4 shows the variation of reserved water height, considering the above-mentioned consumption pattern. As indicated in Figures 6.3 and 6.4, the consumption of the water started from the mid-May and continued until the mid-October.

The thermal variation in vertical direction for some of the selected days, from the beginning to the end of the discharge period, is shown in Figure 6.5. It should be mentioned that the water temperature in different layers was recorded 2 times every 10 days: 8 in the morning and 8 in the evening. The data presented in Figure 6.5 is a selection (sample) of the entire data. There was not any considerable difference between the recorded temperatures in the morning and evening. Therefore, the presented temperature distribution relates to the data recorded in the evening. As observed, from the beginning to the finishing time of water discharge, there had been a stable thermal stratification established during storage period and preserved during the discharge cycle. Two different regions are observed inside the thermally stratified tank during discharge cycle. The bottom region with a linear temperature distribution and the upper one in which a nearly exponential thermal stratification is developed. The deviation of the upper layer from linear temperature profile is due to heat exchange between upper layers of water with roof and evaporative heat and mass transfer induced by the Baudgeers. The reduction of the water height in the tank, during the discharge cycle, is shown in Figure 6.5.

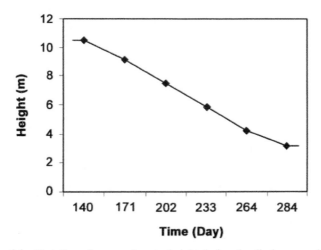

Figure 6.4 Variation of reserved water height during the discharge period [1].

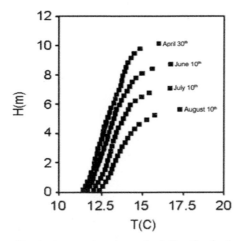

Figure 6.5 Variation of water temperature in vertical direction in the tank in different days of discharge [1].

The variation of the water temperature in radial direction is shown in Table 6.2. Sensors 1 and 2 were placed 1 m away from the walls of the tank and sensor 3 in the center of it. As observed in the table, the distribution of the temperature in radial direction is constant.

The maximum and the minimum variation of the temperature of the upper and lower layers of the water inside the tank and the average ambient temperature during the discharge period are shown in Figure 6.6. The

Table 6.2   Distribution of the temperature in radial direction inside the tank [1]

Date	Sensor 1	Sensor 2	Sensor 3
11/10/2002	15.6	15.6	15.6
11/10/2002	15.4	15.4	15.4
11/10/2002	14.9	14.9	14.9

temperature variation of the lower layer during the discharge time was around 11.5 to 13.1°C, while the average fluctuation of the ambient temperature was between 23 and 38°C. Therefore, all through the 6-month discharge cycle, when the demand for cold water had been high and the ambient temperature had been 42°C, at 4 pm in the hottest day (Figure 6.7), the cold drinkable water had been available for the residents (the temperature of the drinkable water should be between 5 and 15°C and the best temperature is between 8 and 13°C [3]).

As it is seen in Figure 6.6, the rate of temperature variation of the upper layer had been proportional with the ambient temperature changes. The same behavior had been reported by Bahadori [4] too; the changes of the outside ambient temperature had had little effects on the outflow water of the tap.

As it is observed in Figures 6.5 and 6.6, the least temperature is at the level of the outflow water from the tap. The tap of this Aub-anbar was 0.9 m. from the bottom and had a low rate of flow so that there would not be any disturbance in the thermal stratification and it would be kept constant.

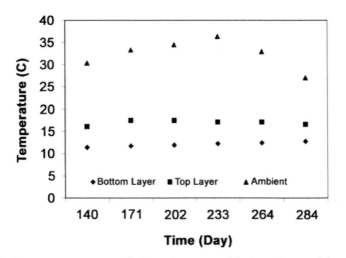

Figure 6.6   Average temperature variation at the upper and the lower layers of the water inside the tank and the average ambient temperature during the discharge period [1].

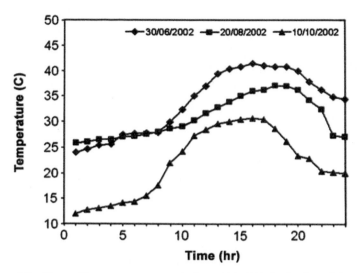

Figure 6.7 The ambient temperature variation within 24 h for 3 selected days [1].

References

[1] Dehghan, A. A., and Dehghani, A. R. (2003). *An Experimental Evaluation of Thermal Stratification in Cisterns, A Research Project, School of Mechanical Engineering*, Yazd University, Yazd, Iran.

[2] Holland, F. (2007). *Fluids Mechanics*, translated by M. Shariati. Hormozgan University, Hormozgan.

[3] Dahlhaus, C., and Damrath, H. (1982). *Wasserversogung*, Abschnitt 3, B. G. Teubner, Stuttgart.

[4] Bahadori, M. N., and Haghighat, F. (1988). Long-term storage of chilled water in cisterns in hot, arid regions. *Building Environ.* 2329–37.

7

Analytical Study of Heat Transfer in Aub-Anbar

In this chapter, the thermal characteristics of Aub-anbars are analytically investigated. The Aub-anbar considered in this study is the same one discussed in Chapter 6. To obtain thermal stratification in the reservoir, some assumptions have to be made, the most important of which is the linearized term of produced radiation because of the heat exchange between the water surface and the internal ceiling of the dome-shaped roof of the Aub-anbar [1–3], which will be discussed in detail in the subsequent parts.

7.1 Analytical Study of Temperature Distribution in the Reservoir

Regarding the fact that heat radiation transfer occurs between the surface water and the dome-shaped roof in Aub-anbar, there is no analytical solution available for this issue. However, assuming that radiation heat transfer coefficient (h_r) is known, a solution can be obtained for the system provided that the boundary condition at the water surface is linearized and we can apply analytical method to solve the problem.

In general, the following assumptions have been made to solve the problem [1, 3]:

1. Because of the symmetry in the angular direction, the reservoir supposed to be two-dimensional and in radial (r) and vertical (z) directions, as shown in Figure 7.1.

2. The soil surrounding the reservoir is homogenous, isotropic, and with constant properties.

3. The effect of the water in the reservoir on the surrounding soil is very little and negligible.

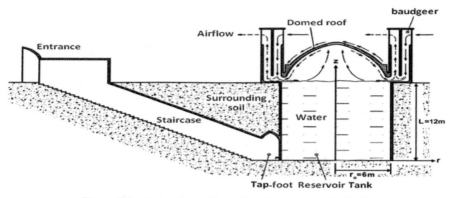

Figure 7.1   A sketches of the Aub-anbar under study [1–3].

4. The heat flow in the regions of the soil adjacent to reservoir has been assumed one-dimensional, with the soil temperature distribution given by [4, 5]:

$$T(z,t) = T_m + \frac{1}{2}A\left[\exp\left(-z\sqrt{\frac{\omega}{2\alpha_s}}\right)\sin\omega\left(t - \frac{z}{\sqrt{2\alpha_s\omega}} + \varphi\right)\right] \quad (7.1)$$

where z is the distance from the ground level and α_s is the thermal diffusivity of the soil. Also, in the above relations, we have

$$T_m = \frac{(T_{\max} + T_{\min})}{2} \quad (7.2)$$

$$A = (T_{\max} - T_{\min}) \quad (7.3)$$

$$\omega = \frac{2\pi}{365}\left(\frac{1}{\text{day}}\right) \quad (7.4)$$

where T_{max} and T_{min} are the mean daily temperatures of the warmest and the coldest months; t is the day number; and ϕ is a time lag. For $\phi = -106$ days, the minimum and maximum temperatures occur on January 15 ($t = 15$) and July 16 ($t = 197$).

5. The heat transfer between the water layers is taking place by conduction. The radiation heat exchange between the water surface and the internal ceiling of the dome-shaped roof of the Aub-anbar is linearized and according to the following relations:

$$q_{\text{rad}} = h_r A(T_S - T_H) \quad (7.5)$$

where T_s is the temperature of the water surface; T_H, the temperature of the internal ceiling of the dome-shaped roof of the Aub-anbar and equal to the mean daily ambient air temperature; and h_r, the radiation heat transfer coefficient.

7.1.1 Governing Equations and the Boundary Conditions

Considering the assumptions made in the previous part, for thermal analysis of water inside the reservoir, governing equations are expressed as follows [1, 3]:

$$\frac{\partial^2 T}{\partial r^2} + \frac{1}{r}\frac{\partial T}{\partial r} + \frac{\partial^2 T}{\partial z^2} = \frac{1}{\alpha_w}\frac{\partial T}{\partial t} \tag{7.6}$$

where α_w is the thermal diffusivity of water. Boundary conditions are equal to [1−3]:

$$T(0, z, t) = \text{finite} \tag{7.7}$$

$$T(r_0, z, t) = T_w \tag{7.8}$$

$$T(r, 0, t) = T_f \tag{7.9}$$

where T_f is the temperature of the floor of the reservoir and T_w is the mean temperature of the reservoir's walls, which are calculated through the Equation (7.1). The boundary condition at the surface of the water is below as well [1−3]:

$$\frac{\partial T}{\partial z}\Big|_{(r,L,t)} = C_1\, T\,(r, L, t) + C_2 \tag{7.10}$$

$$C_1 = \frac{(h_r + h_a)}{k_w} \tag{7.11}$$

$$C_2 = \frac{(\dot{m}_v\, h_{fg} - h_r\, T_H - h_a\, T_a)}{k_w} \tag{7.12}$$

where h_a is the natural convection heat transfer coefficient between water surface and air; h_{fg}, the latent heat of vaporization of water; $\dot{m}_v$, the rate of water evaporation; k_w, the thermal conductivity of water; and T_a, the ambient air temperature.

Computational region and its relevant boundary conditions are depicted in Figure 7.2. Also, the initial condition is equal to

$$T(r, z, 0) = T_i \tag{7.13}$$

where T_i is the initial temperature of water.

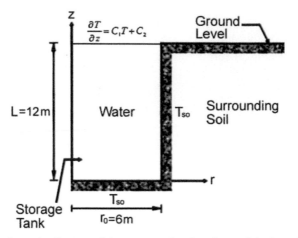

Figure 7.2 A schematic diagram of the computational region and the boundary conditions [1–3].

7.1.2 Solutions of the Governing Equations

To solve the Equation (7.6), regarding the boundary conditions and initial condition, we have to divide it into two steady-state and transient parts and by employing the principle of superposition, obtain the temperature distribution. Therefore, we have [1–3]:

$$T(r, z, t) = \bar{T}(r, z, t) + T_S(r, z) \tag{7.14}$$

The steady-state and transient parts are as follows:

1. *Steady-state part:*

$$\frac{\partial^2 T_S}{\partial r^2} + \frac{1}{r}\frac{\partial T_S}{\partial r} + \frac{\partial^2 T_S}{\partial z^2} = 0 \tag{7.15}$$

$$T_S(0, z) = \text{finite} \tag{7.16}$$

$$T_S(r_0, z) = T_w \tag{7.17}$$

$$T_S(r, 0) = T_f \tag{7.18}$$

$$\frac{\partial T_S}{\partial z}\Big|_{(r, L)} = C_1 T_S(r, L) + C_2 \tag{7.19}$$

2. *Transient part:*

$$\frac{\partial^2 \overline{T}}{\partial r^2} + \frac{1}{r}\frac{\partial \overline{T}}{\partial r} + \frac{\partial^2 \overline{T}}{\partial z^2} = \frac{1}{\alpha_w}\frac{\partial \overline{T}}{\partial t} \tag{7.20}$$

$$\overline{T}(0, z, t) = 0 \tag{7.21}$$

$$\overline{T}(r_0, z, t) = 0 \tag{7.22}$$

$$\overline{T}(r, 0, t) = 0 \tag{7.23}$$

$$\frac{\partial \overline{T}}{\partial z}\Big|_{(r, L, t)} = C_1 \overline{T}(r, L, t) \tag{7.24}$$

$$\overline{T}(r, z, 0) = T_i - T_S(r, z) \tag{7.25}$$

In steady-state part, since all the conditions are non-homogeneous, as in Figure 7.3, first, they should be divided into three parts with each having 3 homogeneous boundary conditions and 1 non-homogeneous boundary

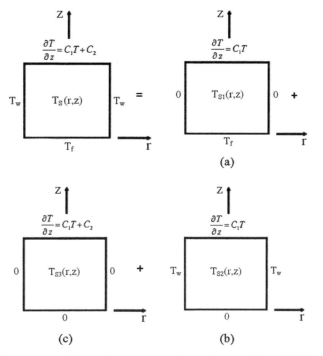

Figure 7.3 Conversion of the non-homogeneous boundary conditions into 3 parts with each having 3 homogenous boundary conditions and 1 non-homogeneous boundary condition [1].

condition. These problems are solved by the method of separation of variables [1–3]. Therefore,

$$T_S(r, z) = T_{S1}(r, z) + T_{S2}(r, z) + T_{S3}(r, z) \qquad (7.26)$$

As a result, by employing the principle of superposition, we can calculate the temperature distribution in the steady-state part as follows [1–3]:

$$T_S(r, z) = \sum_{n=1}^{\infty} \{D_1 \cosh \lambda_n z \, J_0(\lambda_n r) + D_2 \sinh \lambda_n z \, J_0(\lambda_n r)$$
$$+ D_3 \sin \gamma_n z \, I_0(\gamma_n r)\} \qquad (7.27)$$

Then, D_1, D_2, and D_3 are as follows:

$$D_1 = \frac{a_1}{a_2} T_f \qquad (7.28)$$

$$D_2 = \frac{a_1}{a_2} \left(\frac{T_f}{a_3} + a_4 C_2 \right) \qquad (7.29)$$

$$D_3 = a_5 \, a_6 \, a_7 \, T_w \qquad (7.30)$$

$$a_1 = \frac{r_0}{\lambda_n} J_1(\lambda_n r_0) \qquad (7.31)$$

$$a_2 = \frac{r_0^2}{2} \left[\{-\lambda_n J_1(\lambda_n r_0)\}^2 + \{J_0(\lambda_n r_0)\}^2 \right] \qquad (7.32)$$

$$a_3 = \left[\frac{(C_1 \sinh \lambda_n L - \lambda_n \cosh \lambda_n L)}{(\lambda_n \sinh \lambda_n L - C_1 \cosh \lambda_n L)} \right] \qquad (7.33)$$

$$a_4 = [\lambda_n \cosh \lambda_n L - C_1 \sinh \lambda_n L]^{-1} \qquad (7.34)$$

$$a_5 = [I_0(\gamma_n r_0)]^{-1} \qquad (7.35)$$

$$a_6 = \left[\frac{L}{2} - \frac{1}{4\gamma_n} \sin 2\gamma_n L \right]^{-1} \qquad (7.36)$$

$$a_7 = \left[\frac{1}{\gamma_n} (1 - \cos \gamma_n L) \right] \qquad (7.37)$$

where eigenvalue λ_n is the positive roots of the Equation (7.38). Also, the eigenvalue γ_n is calculated through the Equation (7.39).

$$J_0(\lambda_n r_0) = 0 \quad (n = 1, 2, 3, \ldots) \qquad (7.38)$$

$$\tan \gamma_n L = \frac{\gamma_n}{C_1} \quad (n = 1, 2, 3, \ldots) \tag{7.39}$$

In Equation (7.38), J_0 is the Bessel function of first kind of the zero order.

In transient part, the temperature distribution is calculated through the Equation (7.20) by the method of separation of variables and using boundary and initial conditions (7.21) to (7.25) based on the following Equation (7.1)–(7.3):

$$\overline{T}(r, z, t) = \sum_{n=1}^{\infty} B J_0(\lambda_n r) \sin \gamma_n z \, \exp(-\alpha_w(\lambda_n^2 + \gamma_n^2)t) \tag{7.40}$$

Then, B is equal to

$$B = b_1 - a_6(b_2 + b_3 + b_4) \tag{7.41}$$

$$b_1 = \frac{a_1 \, a_6 \, a_7}{a_2} T_i \tag{7.42}$$

$$b_2 = \frac{a_1 \, a_8}{a_2} T_f \tag{7.43}$$

$$b_3 = \frac{a_1 \, a_9}{a_2} \left(\frac{T_f}{a_3} + a_4 \, C_2 \right) \tag{7.44}$$

$$b_4 = \frac{a_5 \, a_7 \, a_{10}}{a_2} T_w \tag{7.45}$$

$$a_8 = \left[\frac{\lambda_n \sin \gamma_n L}{\lambda_n^2 + \gamma_n^2} \sinh \lambda_n L - \frac{\gamma_n \cos \gamma_n L}{\lambda_n^2 + \gamma_n^2} \cosh \lambda_n L + \frac{\gamma_n}{\lambda_n^2 + \gamma_n^2} \right] \tag{7.46}$$

$$a_9 = \left[\frac{\lambda_n \sin \gamma_n L}{\lambda_n^2 + \gamma_n^2} \cosh \lambda_n L - \frac{\gamma_n \cos \gamma_n L}{\lambda_n^2 + \gamma_n^2} \sinh \lambda_n L \right] \tag{7.47}$$

$$a_{10} = \left[\frac{\left(\frac{\gamma_n}{r_0} \right) J_0 \left(\frac{\lambda_n}{r_0} \right) I_0' \left(\frac{\gamma_n}{r_0} \right) - \left(\frac{\lambda_n}{r_0} \right) J_0' \left(\frac{\lambda_n}{r_0} \right) I_0 \left(\frac{\gamma_n}{r_0} \right)}{\lambda_n^2 + \gamma_n^2} \right] \tag{7.48}$$

In Equation (7.48), J_0 is the Bessel function of first kind of the zero order and I_0 is the modified Bessel function of first kind of zero order. J_0' and I_0' are the derivatives of J_0 and I_0, respectively. The eigenvalues λ_n and γ_n are calculated through Equations (7.38) and (7.39). The final temperature distribution in the reservoir is obtained from Equation (7.14).

7.1.3 Selection of the Important Parameters in Determining the Thermal Performance of Aub-Anbars

To obtain the temperature distribution in the reservoir of the Aub-anbar under study, there is a need to select or assume some relevant parameters. These parameters are as follows:

Kind and properties of soil

Since this Aub-anbar was located in the hot and arid region, dry clay soil with 5% moisture content with a density $\rho_s = 1000 \ \text{kg/m}^3$, thermal conductivity $k_s = 0.25 \ \text{W/m}\,^\circ\text{C}$, and a thermal diffusivity $\alpha_s = 2.58 \times 10^{-7}$ m^2/s was used [5].

Properties of water

The properties of water may be assumed with constant properties of thermal conductivity $k_w = 0.56 \ \text{W/m}\,^\circ\text{C}$ thermal diffusivity $\alpha_w = 1.34 \times 10^{-7}$ m^2/s, and latent heat of vaporization $h_{fg} = 2480 \ \text{kJ/kg}$ [5].

Properties of air

Air properties were evaluated at the daily temperature of the ambient air (information obtained from Yazd Meteorological Bureau) [6, 7].

Ambient air conditions

Using the information from Yazd Meteorological Bureau, the average ambient air dew point temperatures, ambient air, and the relative humidity for various months are known. Having average temperature and relative humidity, we can calculate the humidity ratio of the air by using relative humidity chart [6, 7].

Temperature at the water surface (T_S)

Using the experimental data obtained in Chapter 6, we can determine the temperature of the water surface.

Initial temperature of water (T_i)

Using the experimental data, the initial temperature of water was assumed to be 15°C.

Natural convection heat transfer coefficient (h_a)

Assuming some mean value for the airflow from the Baudgeers over the water surface and having the temperature of water surface, we can calculate

h_a for the discharging months. Assuming the airflow from the Baudgeers over the water surface 0.5 m/s, h_a for mid-April to mid-May is 3.8 W/m^2°C and for mid-September to mid-October h_a is 1.25 W/m^2°C [5]. For months between them, the linear variations of h_a were assumed to be between those two values.

Thermal radiation heat transfer coefficient (h_r)

The thermal radiation heat transfer coefficient between the water surface and the internal ceiling of the dome-shaped roof of the Aub-anbar (assumed to be equal to the mean daily temperature of the ambient air) was estimated between 4 and 5 W/m^2°C [5].

Evaporation rate from the water surface ($\dot{m}_v$)

The rate of evaporation from the water surface was determined from the following relation [5]:

$$\dot{m}_v = h_m(W_s - W_a) \tag{7.49}$$

where h_m is the mass transfer coefficient and W_a is the humidity ratio of the air above the water surface and W_s is that if the air was saturated at the water surface temperature. The mass transfer coefficient can be calculated through Lewis relation [5]:

$$h_m = \frac{h_a}{C_{P_a}} \tag{7.50}$$

where C_{pa} is the specific heat of air at constant pressure. The relations between the humidity ratio and other atmospheric parameters are found in air-conditioning texts or ASHRAE Handbooks [8].

Temperature at the floor of the Aub-anbar (T_f)

Using Equation (7.1), we can calculate the temperature at the floor of the reservoir.

Temperature alongside the Aub-anbars walls (T_w)

Temperature distribution in the vertical line may be estimated through Equation (7.1), the average of which is between $x = 0$ and $x = L$ for the boundary condition of the wall (T_w).

It should be reminded that the rate of flow from the tap, located near the bottom of the Aub-anbar, is low. That is why there is no disturbance in the stratification and the distribution of the temperature in the Aub-anbar.

7.2 Presentation of the Results

In order to obtain the results of the analytical analysis, we should solve the Equation (7.14) in the considered month. To do so, in Equations (7.27) and (7.40), only the first serial 100 sentences have been used because of the hyperbolic sine and cosine in the fraction. As it is seen in Figure 7.4, a stable thermal stratification is formed deep down in the Aub-anbar. The same figure also shows the vertical distribution of the temperature in the water in the Aub-anbar, gained in the experimental analysis during the discharge period. Figure 7.4 presents the fact that the graphs obtained in analytical measurements are in good agreement with the ones gained in experimental measurements. The slight difference is due to the estimations of some of the variables in the analytical approach. Similarly, the figure indicates that the stored water can be divided into two separate regions: the bottom region with a linear temperature profile and the upper region where the linear variations are decayed. The deviation of the temperature distribution in upper layers of the water is because of the thermal radiation exchange between the surface layer of the water and the dome-shaped roof of the Aub-anbar, and the convection heat transfer and evaporation caused by the airflow over the water surface via Baudgeers. In fact, it is the stable temperature stratification which makes the cold water to be available through the hot months.

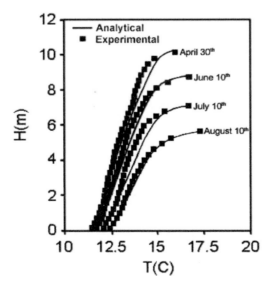

Figure 7.4   Variation of water temperature in vertical direction in the reservoir of Aub-anbar [1–3].

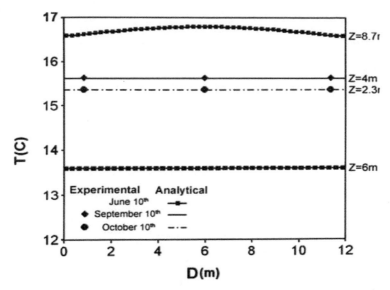

Figure 7.5 Variation of water temperature obtained in analytical measurements and experimental results in radial direction for the different days of the discharge period [1, 3].

In Figure 7.5, the temperature distribution has been shown for the analytical measurements in radial direction, compared to the experimental results. As it is seen, the temperature distribution in radial direction is a fixed value except for the surface water which is curved because of the thermal radiation exchange with the internal ceiling of the dome-shaped roof of the Aub-anbar.

In Figure 7.6, the temperature is shown at the lower and upper water layers in the reservoir together with the average temperature of the outside ambient air during the discharge period for the analytical and experimental results. The same figure indicates that there is a good agreement between the experimental and analytical results and the water is stored satisfactorily cold for drinking in the Aub-anbar all during the consumption period.

There were several parameters affecting the thermal stratification obtained in analytical study; the most important of which includes the following:

1. Dimensions of the reservoir
2. Ambient air temperature
3. The kind and the moisture content of the soil
4. The airflow velocity over the water surface and the rate of water evaporation
5. Thermal radiation transfer coefficient between the water surface and the internal ceiling of the dome-shaped roof of the Aub-anbar

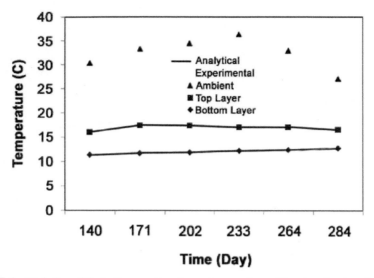

Figure 7.6   Variation of the bottom and top layers temperature of the stored water along with the average ambient air temperature variation during the discharge period [1–3].

6. The initial temperature of the water in the reservoir
7. The rate of water consumption from reservoir during the discharge period.

The parameters mentioned above would be discussed below. Figure 7.7 shows the effect of wind velocity from the Baudgeers on the thermal stratification of water in the Aub-anbar. The temperature variation of water based on the height is shown for different values of induced wind velocity in the 10th of June. Two obtained values were based on the assumptions that the wind velocity was more than the wind velocity of 0.5 m/s for 10 days from 1th to 10th of June. The heat, mass transfer coefficient, and the rate of water evaporation increase due to the increase in wind velocity. Consequently, the water temperature close to the surface decreases with increasing the wind velocity. It was seen that the bottom layers' temperatures were not affected significantly by the induced wind velocity over the water surface. As the wind velocity increases, the rate of evaporation enhances and the water level decreases. The temperature of the layers near the water surface drops as well. It should be noted if the high wind velocity continues for more than 10 days, i.e., beyond 10th of June, it would affect the thermal profile of the lower layers of the water in the Aub-anbar, and in turn, the rate of evaporation increases and the water level drops more.

The effect of variation of the ambient air temperature on the temperature profile along the vertical direction of the stored water in the reservoir is depicted in Figure 7.8. The results for the water temperature variation are

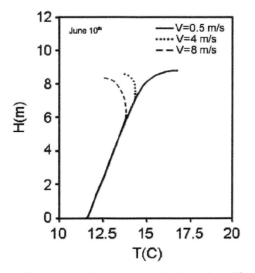

Figure 7.7 The effect of induced wind velocity on the thermal stratification of water inside the reservoir for the 10th of June [1, 3].

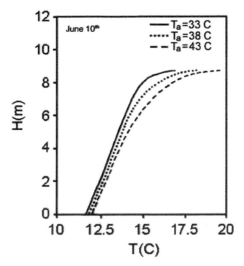

Figure 7.8 The effect of the ambient air temperature on the thermal stratification of water inside the reservoir for the 10th of June [1, 3].

shown for three different ambient air temperatures for the 10th of June. Two values are based on the assumption that the ambienta ir temperature was more than the temperature 33°C, from 1th to 10th of June. When it gets warmer, the water temperature increases. This increase is seen in upper layers which are shown in Figure 7.8. It is also observable that the effects of the ambient air temperature on the lower layers are less. Therefore, the cool stored water in the Aub-anbar is well-preserved against the thermal variation of the ambient air temperature.

The effect of the aspect ratio of the reservoir (the ratio of height to diameter), i.e., $H^* = L/D$, on the thermal stratification is shown in Figure 7.8. The results indicate that the water temperature increases as the ratio of the height to diameter decreases. This increase is less in lower layers. In fact, when $H^* = 1$ and 2, the temperature near the temperature of the water surrounding the tap remains almost constant so that if the aspect ratio becomes less than 1, the water near the tap would become warmer. For a certain capacity of the reservoir, an increase in aspect ratio, i.e., the increase in the ratio of height to diameter, is an indication of the better thermal performance, but digging and constructing of the reservoir becomes more difficult. Also, for a certain capacity of the reservoir, the decrease in the aspect ratio, i.e., decrease in the ratio of height to diameter, is an indication of a worse thermal performance, but the digging and the construction of the reservoir is easier. Therefore, 1 is the best aspect ratio in this type of reservoirs. This shows that Iranian had been quite aware of both the construct and the thermal performance in creating the masterpiece of Aub-anbar (Figure 7.9).

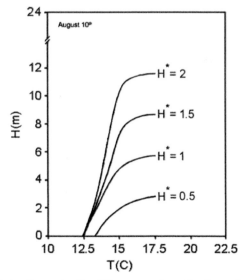

Figure 7.9   The effects of aspect ratio of the reservoir on the thermal stratification of water inside the reservoir for the 10th of August [1, 3].

References

[1] Dehghani, A. R. (2006). *Analytical and Numerical Investigation of Heat Transfer in Cisterns,* MS Thesis, Department of Mechanical Engineering, Islamic Azad University, Science and Research Branch, Tehran, Iran.

[2] Arefmanesh, A., Dehghan, A. A., and Dehghani, A. R. (2009). *Modeling Heat Transfer in Cisterns: Experimental and Theoretical Studies,* Applied Thermal Engineering, 29, 3261–3265, Oct.

[3] Dehghan, A. A., and Dehghani, A. R. (2011). Experimental and Theoretical Investigation of Thermal Performance of Underground Cold-water Reservoirs. *Intl. J. Thermal Sci.* 50(5), 816–824, May.

[4] Ozisik, M. N. (2000). *Heat Conduction.* John Wiley, New York.

[5] Bahadori, M. N., and Haghighat, F. (1988). Long-term storage of chilled water in cisterns in hot, arid regions, *Building Environ.* 23, 29–37.

[6] Dehghan, A. A., and Dehghani, A. R. (2003). *An experimental evaluation of thermal stratification in cisterns.* A Research Project, School of Mechanical Engineering, Yazd University, Yazd, Iran.

[7] Dehghan, A. A. (2004). "An Experimental Investigation of Thermal Stratification in an Underground Water Reservoir," in *Proceeding of HT-FED 2004-56784: 2004 ASME Heat Transfer/Fluids Engineering summer conference Westin Charlotte and Convention Center.* Charlotte, North Carolina, USA, 1–4, July 11–15.

[8] ASHRAE Handbook. (1981). *1981 Fundamentals, American Section of Heating*, Refrigerating and Air-Conditioning Engineers, Inc., Atlanta, Georgia.

8

Numerical Analysis of Heat Transfer in Aub-Anbars: Finite Difference Method

The exact thermal analysis of Aub-anbars is a formidable task. Bahadori and Haghighat have carried out the numerical investigation of the heat transfer in Aub-anbars by the finite difference method [1]. In that study, the term of produced radiation because of the heat exchange between the water surface and the internal ceiling of the dome-shaped roof of the Aub-anbar was considered linear [1–3]. To obtain thermal stratification in the water inside the reservoir and the surrounding soil, some simple assumptions were considered which will be discussed in detail in the subsequent sections.

8.1 Assumptions

Since the investigation of thermal performance in Aub-anbars is extremely a difficult task and in order to simplify the analysis and still be able to obtain reasonably accurate results, an Aub-anbar with a square floor, 16×16 m to the side, and the height of 8 m was chosen, and the following assumptions were made [1, 2].

(1) A sinuses function can be considered for the variations of the daily temperature which is maximum at 3 pm and minimum at 3 am, and for the mean daily temperature of the ambient air during the year, a sinuses function which can be represented by the relations Equations (8.1) through (8.4) which is maximum on June 16 and minimum on January 15.

$$T = T_m + \frac{1}{2}A \sin \omega (t + \varphi) \tag{8.1}$$

$$T_m = \frac{(T_{\max} + T_{\min})}{2} \tag{8.2}$$

$$A = (T_{\max} - T_{\min}) \tag{8.3}$$

$$\omega = \frac{2\pi}{365}\left(\frac{1}{\text{day}}\right) \qquad (8.4)$$

where $T_{\max}$ and $T_{\min}$ are the mean daily temperatures in July (the warmest month) and January (the coldest month), respectively, and t is the day number and ϕ in this relation is a time lag. For $\phi = -106$ days, the minimum and maximum temperatures occur on January 15 ($t = 15$) and July 16 ($t = 197$), respectively.

(2) The soil temperature at different points is specified as a result of the heat exchange between the soil and the environment. This heat exchange occurs through the sun shine, heat radiation to atmosphere, convection heat exchange with the environment, water evaporation from surface, and conduction heat in the soil.

To simplify the investigation, it is assumed that the daily soil temperature 0.5 m, below the ground surface, is equal to the mean of the temperature on the specified day. If we take that depth (0.5 m) as the reference point, the temperature variation for different depths can be calculated through the following relation [3]:

$$T(y, t) = T_m + \frac{1}{2}A\left[\exp\left(-y\sqrt{\frac{\omega}{2\alpha_s}}\right)\sin\omega\left(t - \frac{y}{\sqrt{2\alpha_s\omega}} + \varphi\right)\right], \quad (8.5)$$

where y is the distance from the ground level and α_s is the thermal diffusivity of the soil.

(3) Water temperature from the bottom to the surface varies in reservoirs, but this thermal stratification does not change as the water is discharged from the bottom (with maximum discharge of 30 l/min).

(4) The soil surrounding the Aub-anbar is homogeneous, isotropic, and with constant properties. This assumption, although far from reality, is necessary to obtain a general result for the performance of Aub-anbars. Otherwise, one has to determine the soil properties of a site and then analyze the performance of the Aub-anbar to be built in that location.

(5) The temperatures of the soil at vertical surfaces 10 m away from each wall and a horizontal surface 10 m below the floor of the Aub-anbar (18 m from the ground level) are unaffected by the presence of the Aub-anbar. That is, water in the Aub-anbar with a volume of 2048 m³ is interacting with its surrounding soil with a volume of 21,280 m³.

(6) The heat flow in the regions of the undisturbed soil surrounding the Aub-anbar (at distances of $x = 10$ m or more from the Aub-anbar walls) is one-dimensional, with the soil temperature distribution given by relation (8.5) above and $y = 18$ m. This temperature is constant through the entire year [15].

(7) The heat flow in the soil surrounding it at distances 10 m away from the walls and the floor and in the water of the Aub-anbar 2 m away from the walls is two-dimensional. The heat transfer between the water layers is taking place by conduction and between the surfaces (walls and floor) and water by natural convection.

(8) The heat flow in water at distances of 2 m or more from the wall is one-dimensional. The water temperature at each level is equal to the water temperature at the same level 2 m from the wall.

(9) The water surface in the Aub-anbar when it is full is 0.5 m below the ground level, and the temperature of the surrounding soil at this level remains equal to the mean daily ambient air temperature of that day.

(10) Assuming that the heat flow is two-dimensional, one can assume a section of the Aub-anbar with 1 m width and write the relations governing that section and then apply the obtained results for the water and soil temperature to the real Aub-anbar.

8.2 Development of the Governing Equations

Regarding the above assumptions, one can assume a hypothetical heat network for the water and surrounding soil of Aub-anbar and write the first thermodynamics law for its several nodes and do those relations through a two-dimensional finite difference approximation. Figure 8.1 shows a section of the Aub-anbar and its surrounding soil with thickness of 1 m (vertical dimension to the plane of the paper) and considered nodes [1, 2].

The grid sizes of 1 m near the wall and the floor of the Aub-anbar, then of 2, and finally of 4 m at distances far from the Aub-anbar wall and floor are chosen. For nodes A to I which have different conditions, the first law (the algebraic sum of the transferred heat to a node is equal to the internal energy variations in unit of time for the same node) is as follows:

(1) For node A,

$$T'_{i,j} = T_{i,j} \left[1 - \alpha_s \, \Delta t \left(\frac{1}{2xx_r} + \frac{1}{2xx_l} + \frac{1}{2zz_b} + \frac{1}{2zz_t} \right) \right]$$
$$+ \alpha_s \, \Delta t \left[\frac{T_{i+1,j}}{2xx_r} + \frac{T_{i-1,j}}{2xx_l} + \frac{T_{i,j+1}}{2zz_b} + \frac{T_{i,j-1}}{2zz_t} \right], \qquad (8.6)$$

where α_s is the thermal diffusivity of the soil, Δt is the time interval, T and T' are the temperatures of the soil at the present and the next time step,

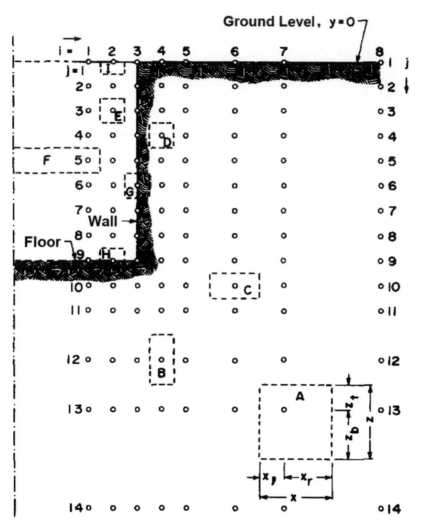

Figure 8.1   A section of the Aub-anbar and its surrounding soil considered for thermal analysis. The nodes are spaced, depending on their locations, 1, 2, or 4 m from each other [1].

respectively, and x, z, x_r, x_1, z_b, and z_t are the dimensions shown in Figure 8.1, and i refers to columns and j to the rows of nodes. Referring to this figure, for the node A, $x = z = 3$ m and $x_1 = z_t = 1$ m and $x_r = z_b = 2$ m.

(2) For nodes such B, C, and D, *where* $x_r = x_1 = 1/2x$ and $z_t = z_b = 1/2$ m, the above relation simplifies to the following equation:

$$T'_{i,j} = T_{i,j} \left[1 - 2\alpha_s \, \Delta t \left(\frac{1}{x^2} + \frac{1}{z^2} \right) \right]$$

$$+ \, \alpha_s \, \Delta t \left[\frac{(T_{i+1,j} + T_{i-1,j})}{x^2} + \frac{(T_{i,j+1} + T_{i,j-1})}{z^2} \right]. \tag{8.7}$$

For node B, $x = 1$ m and $z = 2$ m; for node C, $x = 2$ m and $z = 1$ m; and for node D, $x = z = 1$ m. The above explicit finite-difference equation is stable if the following relation is satisfied [3]:

$$1 - 2\alpha_s \, \Delta t \left(\frac{1}{x^2} + \frac{1}{z^2} \right) \rangle 1. \tag{8.8}$$

For the values of x and z, the time steps of one and a half hour to one day, and the soil properties considered in this study, the condition of Equation (8.8) is always satisfied [1].

Equations such as (8.6) and (8.7) were written for all the nodes which are in columns 4–7 and rows 2–13, as well as column 13 and rows 10–13 (a total of 60 nodes). For the nodes on the ground level or the ones with columns 4–8 and row 1 which constitute one of the boundary conditions, the temperature of every day is determined from Equation (8.1).

For all the nodes in columns 1–8 and row 14, which constitute one of the boundary conditions, the temperatures are determined from Equation (8.5) where $y = 18$ m. For the nodes which are in column 8 and rows from 1 to 14, temperatures are determined from Equation (8.5) by substituting $j = 1$ for y value in the equation. As can be seen from these equations, the boundary conditions are not constant, but variable with time.

(3) For the nodes such as E located in the water near the wall of the Aub-anbar, the following relation can be written as follows:

$$T'_{2,j} = T_{2,j} \left(1 - \frac{h_w \Delta t}{x C_w} - \frac{\alpha_w \Delta t}{x^2} - \frac{2\alpha_w \Delta t}{z^2} \right) + \frac{h_w \Delta t T_{3,j}}{x C_w}$$

$$+ \, \frac{\alpha_w \Delta t T_{1,j}}{x^2} + \frac{\alpha_w \Delta t (T_{2,j+1} + T_{2,j-1})}{z^2} \tag{8.9}$$

where $x = z = 1$ m and h_w is the convection heat transfer coefficient between the wall of Aub-anbar and water, C_w is the thermal capacity of the water in $\text{kJ/m}^{3\circ}\text{C}$, and α_w is the thermal diffusivity of water in m^2/s. In this equation, we have considered a convocation heat transfer between the node E and the wall adjacent to it, and conduction heat transfer between this node and the nodes surrounding it in the water.

Similar relations can be written for all nodes with $i = 2$ and $j = 2.8$. For nodes with $j = 8$ and 9, convection from the floor was considered.

(4) For node F, we have

$$T'_{1,j} = T_{1,j}\left[1 - \alpha_w\,\Delta t\left(\frac{1}{2x_r(x_r + x_l)} + \frac{2}{z^2}\right)\right]$$

$$+ \alpha_w\,\Delta t\left[\frac{T_{2,j}}{2x_r(x_r + x_l)} + \frac{(T_{1,j+1} + T_{1,j-1})}{z^2}\right] \tag{8.10}$$

where $x_r = 1/2$ m, $x_1 = 6$ m, and $z = 1$ m. Similar relations can be written for all similar nodes with $i = 1$ and $j = 2.8$.

(5) For node G, which represents a point on the wall of the Aub-anbar, we have the following relation:

$$T'_{3,j} = T_{3,j}\left[1 - \frac{2k_s\Delta t}{(C_s + C_w)x^2} - \frac{2h_w\Delta t}{(C_s + C_w)x} - \frac{2(k_s + k_w)\Delta t}{(C_s + C_w)z^2}\right]$$

$$+ \frac{2k_s\Delta t T_{4,j}}{(C_s + C_w)x^2} + \frac{2h_w\Delta t T_{2,j}}{(C_s + C_w)x} + \frac{(k_s + k_w)\Delta t(T_{3,j+1} + T_{3,j-1})}{(C_s + C_w)z^2} \tag{8.11}$$

Where k_s and C_s are the thermal conductivity and the thermal capacity of the soil, respectively, k_w and C_w are the same terms for the water. In this equation, $x = z = 1$ m. Similar relations can be written for all the nodes on the wall of the Aub-anbar ($i = 3$) and $j = 2.8$.

(6) For node H which is a point on the floor of the Aub-anbar, we have a relation similar to that given by Equation (8.11), written for the node G, except that we have to use h_f, or the convection heat transfer coefficient (instead of h_w). The equations for other nodes on the floor are similar to what have been written for the node H. For the node at the corner ($j = 9$ and $i = 3$), there is heat transfer by conduction from the soil on three sides and convection from the water side.

(7) For node I on the water surface, we have the following:

$$T'_{2,1} = T_{2,1}\left[1 - \frac{2(h_a + h_r)\Delta t}{C_w z} - \frac{\alpha_w\,\Delta t}{\left(\frac{2}{z^2} + \frac{1}{x^2}\right)} - \frac{h_w\Delta t}{C_w x}\right] + \frac{2(h_a + h_r)\Delta t T_a}{C_w z}$$

$$+ \alpha_w\,\Delta t\left(\frac{2T_{2,2}}{z^2} + \frac{T_{1,1}}{x^2}\right) + \frac{h_w\Delta t T_{3,1}}{C_w x} + \frac{2\dot{m}_v h_{fg}\Delta t}{C_w z} \tag{8.12}$$

where h_a is the convection heat transfer coefficient between water surface and the air on top, h_r is the thermal radiation heat transfer coefficient between the water surface and the ceiling, $\dot{m}_v$ is the rate of evaporation from the water's surface, and h_{fg} is the latent heat of vaporization of water.

Similar equations were written for the other node on the water surface and for the node at the top corner of Aub-anbar. For the node at the top corner ($i = 3$ and $j = 1$), we considered heat and mass transfer from the water surface (similar to node I) and heat transfer from the node on its right (through the soil) by conduction, its left (through water) by convection, and through the soil and water from the node underneath (similar to node G) by conduction.

When water level drops due to the removal of water from the bottom of the storage, the walls of the Aub-anbar become exposed to the air in the Aub-anbar. They then exchange heat with the air by convection and with the ceiling by thermal radiation. The same is true when water is completely withdrawn from the Aub-anbar (for cleaning purposes) and it is left empty. The heat transfer equations of the nodes on the wall and the floor were modified as these surfaces became in contact with the air. The equations of the nodes have been written for these conditions [1, 2].

8.3 Selection of the Important Relevant Parameters

To solve the above governing equations and obtain the temperature of water in Aub-anbar and the soil surrounding it, we need to select or assume several parameters. In selecting or assuming these parameters, we had to keep in mind the approximate nature of the analysis, the computational costs, and the fact that the exact description of these parameters does not necessarily improve the accuracy of the results greatly.

Of particular importance was the consideration of the soil properties which we assumed remaining constant. The soil surrounding Aub-anbar is never homogeneous nor with constant properties. With a rainfall and water migration in the soil, the properties change both spatially and with time. Yet, we had to assume constant properties for the soil (as well as two-dimensional heat flow) in order to simplify the analysis, save on the high computational costs, and be able to estimate the temperatures of water and the soil.

Selection of the use pattern of the potable water

To simplify our analysis, we assumed a constant water consumption rate from Aub-anbar beginning mid-April and ending mid-December. In other time of the year, Aub-anbar remained full. During each succeeding month, the water level in Aub-anbar drops about 1 m.

8.3.1 Selection of the Soil Properties

We considered two types of soil: (1) a dry clay soil with 5% moisture content with a density, $\rho_s = 1000 \text{ kg/m}^3$, thermal conductivity, $k_s = 0.25 \text{ W/m}°\text{C}$, and a thermal diffusivity, $\alpha_s = 2.5 \times 10^{-7} \text{m}^2/\text{s}$; (2) a dry sandy soil with 5% moisture content with a density, $\rho_s = 1600 \text{ kg/m}^3$, thermal conductivity, $k_s = 1.25 \text{ W/m}°\text{C}$, and thermal diffusivity, $\alpha_s = 8.05 \times 10^{-7} \text{m}^2/\text{s}$. These two types of soil were selected to provide rather extreme conditions for dry soils which may be found in the hot, arid regions.

8.3.2 Selection of Water and Air Properties

We assumed water with constant properties of: thermal conductivity, $k_w = 0.56 \text{ W/m}°\text{C}$, thermal diffusivity, $\alpha_w = 1.34 \times 10^{-7} \text{m}^2/\text{s}$, and latent heat of vaporization, $h_{fg} = 2480 \text{ kJ/kg}$. The air properties were evaluated at the mean daily temperature of the ambient air.

8.3.3 The Ambient Air Conditions

We assumed the potable water storage system to be located in a hot and rather arid region with daily temperatures fluctuating between 20 and 40°C in July and between 0 and 10°C in January, which correspond with the mean daily temperatures in July, T_{max}, and in January, T_{min}, of 30 and 5°C, respectively. It is clear that the hourly temperature variations do not affect the operation of the storage greatly and are neglected in our analysis.

The average ambient air dew point temperatures for different months were selected between 45% in January and 15% in July. Having $T_{max} = 30°\text{C}$ and $T_{min} = 5°\text{C}$, the average ambient air temperature may be calculated by Equation (8.1). By selecting the average relative humidity in each month, the average ambient air dew point temperature can be (referring to the humidity measuring plot) determined [1, 2].

Evaluation of the convection heat transfer coefficients, h_w and h_f

With the operating conditions of the Aub-anbar considered in this analysis, the Rayleigh number is between 10^{11} and 10^{13}, and the natural convection heat transfer between the Aub-anbar's walls and the floor with water is turbulent. Therefore, the heat transfer coefficients h_w and h_f can be determined from the following relations: [1, 2, 4].

$$h_w = 76.5(\Delta T_w)^{\frac{1}{3}} \tag{8.13}$$

$$h_f = 114.75(\Delta T_f)^{\frac{1}{3}} \tag{8.14}$$

where T_w is the difference of temperature between the Aub-anbar's walls and the water and T_f is that of Aub-anbar's floor and the water.

8.3.4 Evaluation of the Convection Heat Transfer Coefficients, h_a and h_r

Through trial-and-error process, it was found that the water temperature is always lower than the air temperature except for the month of November, when the water level in the Aub-anbar is at its minimum. The natural convection heat transfer coefficient, h_a, between water and the air in October was estimated to be 1.25 W/m²°C, when the Aub-anbar is full of water (assumed to be during the months of January through April), the Baudgeers circulate air over the water surface with an average speed estimated to be 0.5 m/s. In this case, the convection heat transfer coefficient was estimated to be $h_a = 4.3$ W/m²°C. For the months of April through October, we assumed a linear variation of h_a between the above two values [1, 2]. When the walls of the Aub-anbar are exposed to the air, the natural convection heat transfer coefficient was estimated to be 3.35 W/m²°C. In order to estimate the variable h_r, refer to Chapter 7, Section 7.1.3.

8.3.5 Evaporation Rate from the Water Surface

In order to estimate the variables h_m and $\dot{m}_v$, refer to Chapter 7, Section 7.1.3.

8.4 Solving the Equations and Determining the Temperature of the Water and the Soil

Regarding the assumptions and determining the mean temperature for all the days of each month and estimation of heat transfer coefficients h_f, h_w, h_a, h_r and the rate of water evaporation $\dot{m}_v$ for the conditions of each month, all written energy equations for different nodes were solved simultaneously and the temperature of the water and the soil were determined for the nodes.

The time interval Δt chosen for the first 5 days of operation of Aub-anbar was 30 min and since then, 1 day. Figure 8.2 shows the temperature variations for the clay and sand soil in the first year of operation of Aub-anbar, and Figure 8.3 shows the temperature of the water at the bottom of Aub-anbar (around the discharge location) for different years of operation for dry clay and dry sandy soils.

In the first year of operation of Aub-anbar, the water is in touch with the relatively warm surrounding water and its temperature is rather high, but

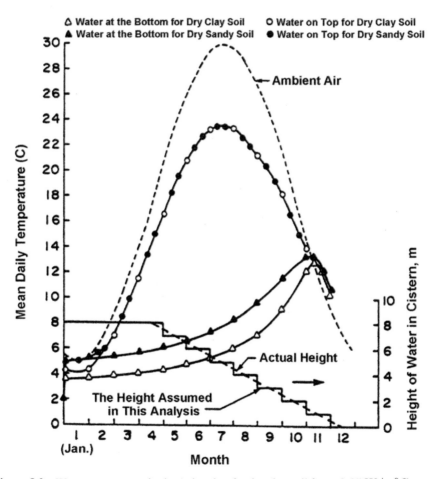

Figure 8.2   Water temperature in the Aub-anbar for dry clay soil $k_s = 0.25$ W/m°C, $\alpha_s = 2.58 \times 10^{-7}$ m^2/s and dry sandy soil $k_s = 1.25 \frac{W}{m°C}$, $\alpha_s = 8.05 \times 10^{-7} \frac{m^2}{s}$ with initial water temperature of 2°C for the first year of operation [1, 2].

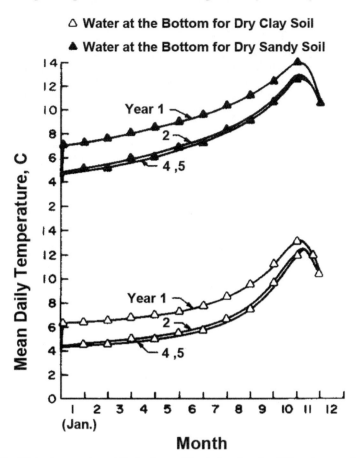

Figure 8.3   Water temperature at the bottom of the Aub-anbar for different years of operation for dry clay and dry sandy soils with the same properties mentioned in Figure 8.2. The initial water temperature is 4°C [1, 2].

gradually as the time passes and through the succeeding years, as the temperature of the soil drops, the temperature of the water also drops accordingly. After the first 4 years of operation of Aub-anbar, water takes a fixed trend in its temperature variations (Figure 8.3).

It should be reminded that the accurate investigation of the thermal performance of Aub-anbars is a difficult task and needs a meticulous effort in doing the estimations. In the present study, the researchers have tried to simplify the investigation by assuming some assumptions and make the task as easy as possible.

References

[1] Bahadori, M. N., and Haghighat, F. (1988). Long-term storage of chilled water in cisterns in hot, arid regions. *Building and Environment* 23, 29–37.

[2] Bahadori, M. N., and Yaghobi, M. (2006). *Natural Cooling and Air-Conditioning in Traditional Buildings in Iran.* Nashr-e Danshgahi Publications, Tehran.

[3] Ozisik, M. N. (2000). *Heat Conduction.* (New York: John Wiley).

[4] Incropera, F. P., and DeWitt, D. P. (1985). *Introduction to Heat Transfer.* New York John Wiley, New York.

9

Modeling Heat Transfer in Aub-Anbars by Artificial Neural Networks (ANN)

In recent years, there has been a ceaseless effort of moving from theoretical research to practical studies, particularly in the field of data processing. These practical studies are helpful for solving the problems that seem to be unsolvable or difficult to solve. Also, there has been an increasing interest in the development of dynamic theoretical systems of model-free, which are based on experimental data. "Artificial neural networks" (ANN) are among those dynamic systems that can transfer the hidden laws beyond the data to the network structure by processing the experimental data. These types of dynamic systems are called intelligent systems, because they can learn the general laws of data based on the calculation of numerical data or samples [1].

9.1 Concept of Neural Networks

A "neural network" is a collection of bio-neurons linked together. It is referred to artificial neural networks which are made of artificial neurons. Therefore, the phrase of "neural network" refers to the two following concepts:

1. Bio-neural network
2. Artificial neural network.

Since 1911, the efforts to understand brain's function and capabilities have been increased. At that time, Segal for the first time reported that the brain is made up of the elements called neurons [1].

Each biological neuron is as complex as microprocessor, but does not have the microprocessor processing speed. Some parts of the biological neurons are created in the birth time, and some other parts are developed in the lifetime, particularly in the early years of life. Scientists in the field of biology have recently found out that the functions of data storing and saving of the

165

neurons occur in the neurons and in the linkage between them [1]. Technically speaking, "learning" is known as creating new connections among neurons and adjustments of existing ones.

It should be noted that the artificial neural networks of a computer application or semiconductor chips have an extremely higher speed (a million times faster) than bio-neurons. However, they carry only a fraction of their capabilities.

Artificial neural networks are progressing in the aspects of analytical capabilities, structural development, and hardware requirements. At present, neural networks have wide applications in scientific activities and technical-engineering issues such as control systems, processing of signals, and identification of patterns.

9.2 Definition and History of Neural Networks

Artificial neural networks are composed of simplified models of biological neurons which are connected to each other through weight coefficients. Several definitions have been proposed for the artificial neural network and one of those is as follows [2–5]: "A processing system which is composed of many well connected simple processors (neurons) that its structure is based on the brain's neural networks' structure." It should be noted the study of neural networks commenced in late 19th and early 20th centuries. In this period, fundamental research was done in physics, psychology, and neuro-physiology by scientists such as Hermann Von Helmholtz, Ernst Mach, and Ivan Pavlov [1]. The primary research mostly emphasized on general learning theories, vision, and conditions and never considers the specific mathematical models of neurons' performance.

A new perspective of neural networks started in 1940s by Warren Mc Culloch and Walter Pitts. They showed that ANN can solve arithmetic and logical functions. Nowadays, their study is usually considered as a starting point for this field of research. This field of research continued with the works of Donald Hebb. He defined the classical conditioning that was proposed by Pavlov as a neurons' characteristic and proposed a new learning mechanism of biological neurons [1].

The earliest practical application of neural networks was in the late 1950s, when Frank Rosenblalt introduced Perceptron network in 1958. Rosenblatt and his colleagues created a network which was able to distinguish different patterns. In 1960, Bernurd Widrow suggested the neural network of adaptive linear element (ADALINE) with new learning rules which was similar to perceptron network in terms of structure [1].

Both of these networks, perspetron and ADALINE, could just categorize the patterns that were linearly distinguishable. Widrow and Rosenblatt knew this limitation and proposed a learning rule for single-layer neural networks; however, they could not improve the learning algorithm of one-layer networks.

Neural networks developed more in 1970s. In 1972, Teo Kohonon and James Anderson independently and unaware of each other introduced new neural networks, which were able to be "saving elements." Stefan Grossberg was working on self-organizing networks in this decade [1]. Neural network progress was slow in 1960s in comparison with 1980s due to the lack of new ideas and speedy computers. In 1980s, the microprocessor technology had more progress, more research was done on neural networks, and new ideas were proposed. They were enough for another good progress in the field of neural networks. In this progress, two aspects should be considered. First, the use of accidental mechanism that mostly explains the performance of feedback (recurrent) networks that were used for data saving. This idea was proposed by an American physicist, John Hopfield, in 1982. The second idea which played a key role for the development of neural networks in 1980s is the "error back-propagation" algorithm that was proposed by David Rummelhart and James Mcvland in 1986. Those two ideas made a good progress in neural networks [1].

In recent years, thousands of papers have been written on the neural networks and their applications. Neural networks have wide applications in different scientific fields. They have a good theoretical and practical progress, but this progress was not in a certain manner and was slow in some periods. The most progress in neural networks relates to modern structures and new learning methods. Generally, neural networks cannot generate a relation between two sets of numbers as a math function, but they can generate a transformation between an input–output series. This transformation gives acceptable outputs with specific accuracy for continuous inputs.

In recent years, neural networks are widely used in heat transfer problems [6–10]. When there is lack of information about internal processes and boundary conditions of a heat transfer problem, using neural network is a proper method for modeling heat transfer [11–13].

In this chapter, the thermal performance of Aub-anbars is investigated by artificial neural networks. The Aub-anbar under consideration is the same Aub-anbar discussed in Chapter 6.

9.3 Architecture and Training Algorithm of the Artificial Neural Networks[1]

9.3.1 Multilayer Perceptron Neural Network (*MLP*) by a Back-Propagation Algorithm (*BP*)

In this study, the multilayer perceptron network by a training back-propagation algorithm has been used. In this network, the number of neurons in the input layer is defined according to the problem. For the current problem, two cases are considered as the inputs of the network. In the first case, only the outside ambient temperature is used as the input. In the second case, the date and the sensors' number that show the level of different points from the bottom of the Aub-anbar are used in addition to the ambient temperature [2–5].

In the first case, the number of neurons in the output layer is 36, which shows the temperature in 36 points of the Aub-anbar. Points are located in different heights within 30 cm vertical distance from each other. In the second case, the number of the neuron in the output layer is one and shows the temperature at the specific level (height) of the Aub-anbar on a specific date. The number of neurons in the internal layers cannot be determined like the input and output layers. This parameter and even the number of internal layers of the network are determined using trial-and-error process to obtain the best architecture of the network.

The following equation describes network behavior:

$$a^0 = P$$
$$a^{L+1} = f^{L+1}\left(w^{L+1} a^L + b^{L+1}\right), \quad L = 0, 1, 2, \ldots, L - 1. \tag{9.1}$$

In this equation, P is the main input of the network and is the input vector of the first layer. Also, L is the number of layers in the network. At first, the input vector (P) is multiplied by weight function (w). The result is added to Bias (b). Finally, the result is applied to the transfer function (f) as an input to make the final result.

The transfer function (f) is usually a gradient or a sigmoid function. Bias (b) and weight (w) are two calibration parameters in neurons. The input vector P with $P_1, P_2, \ldots$ to P_R elements and weight vector w with $w_{1,1}, w_{1,2}, \ldots$ to $w_{1,R}$ elements have been specified. In order to apply weight values to input values, two vectors of P and w should be multiplied by each other in a matrix fashion (Figure 9.1). Then, the bias number has been added to its result. The

---

[1]For more information about artificial neural networks, refer to reference books in this field.

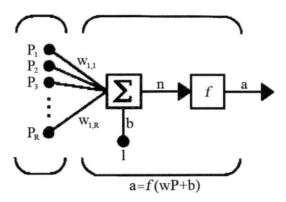

Figure 9.1 A neuron with an input vector [14].

final result (n) is the input of the transfer function (f) that is calculated through Equation (9.2).

$$n = w_{1,1} P_1 + w_{1,2} P_2 + \ldots + w_{1,R} P_R + b \qquad (9.2)$$

The output of the last layer is the output vector of the network. It means that the answer of the network is equal to a^L.

This network is a supervised network. It means that there is an output for each input. The learning data pairs in *BP* network are as follows:

$$(P^1, t^1), (P^2, t^2), \ldots, (P^i, t^i) \qquad (9.3)$$

In this equation, p^i is the input vector of network and t^i is the target vector for input p^i. In the case of Aub-anbars, t^i is the experimental data that was used to train network.

Error signal, in the neurons of output layer (layer (L)) in k^{th} frequent, is obtained from the following equation:

$$e_j(k) = t_j(k) - a_j(k) \qquad (9.4)$$

The only observable outputs are the outputs of the last layer; therefore, the error signal just considers these outputs and compares those with the target vector (experimental data). In the Equation (9.4), $t_j(k)$ is the Jth element of the output vector related with the input vector $P(k)$.

The behavior of a network is evaluated by performance function:

$$F(k) = \sum_{j=1}^{S^L} e_j^2(k) \qquad (9.5)$$

In this equation, S^L is the number of neurons in the layer L.

The optimized answer in *BP* algorithm is specified using sum of squared error between experimental data and the network outputs.

Generally, pairs of data (inputs and target vectors) are divided into three sets: training, evaluation, and test. Usually, sixty percent are devoted to training set, twenty percent to evaluation set, and twenty percent to test set. The neural network's parameters such as weights and biases are calibrated by training set. The evaluation set is used to validate obtained weights and biases. Finally, the obtained network is tested by test set to find out the accuracy of network's output in confronting the inputs which had never been applied to it before and compare the result with the target vector (experimental data).

One of the most important factors in network training is the number of epochs that the network performs during the training process. In the training process of a network, as the number of epochs increase, the difference between the target and output of the network decreases. However, by excessively increase in the number of epochs, the error of the test set increases too. The best number of training epochs is calculated by considering the lowest error in test, evaluation, and training sets [2–5].

9.3.2 Data Preparation

Data should be prepared before the process. In neural networks, the data preparation is not as complex as statistical models. The main point is changing data scales and locating them in the domain of activation function. The tangent-sigmoid activation function is used in the hidden layer of the network. The limitation of this function is between 1 and –1. Also, the slope of this function is considerable at around $n = 1$ and $n = -1$.

In this research, to prevent network saturation, all data change scales have been located between 1 and –1. Figure 9.2 shows the tangent-sigmoid function and its derivative with respect to n. Tangent sigmoid function and its derivative are as follows:

$$a = f(n) = \frac{2}{1 + e^{-2n}} - 1 \tag{9.6}$$

$$a' = f'(n) = 1 - (f(n))^2 \tag{9.7}$$

9.3.3 Calibration Parameters of MLP Artificial Neural Network

Calibration parameters of this type of neural network include the number of hidden layers, transfer functions of each layer, number of neurons in each hidden layer, initial weights of the network, and learning rate.

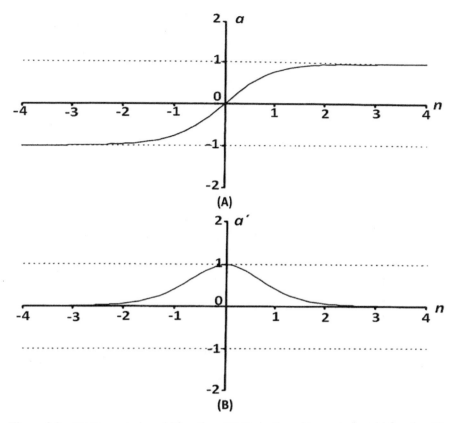

Figure 9.2 (A) Tangent-sigmoid function; (B) Derivative of tangent-sigmoid function [5].

In this research, a *MLP* artificial neural network with one hidden layer is used. The hidden layer's transfer function is tangent-sigmoid and the output layer's transfer function is linear. The number of neurons in the hidden layer is determined with a trial-and-error process. The initial weights of the network are determined by random numbers with unique distribution. Also, the behavior of variables and their correlation during the period may be changed; thus, a fixed learning rate may not be optimized. Therefore, the learning rate is considered to be variable with time. For this purpose, the rate of performance function in each epoch is used, but the value of learning rate is limited to 10^{-2}, and then, the maximum performance function is defined. The maximum performance function (sum of squared errors between the modeling and experimental values) is the highest acceptable limit of the performance function that should be satisfied in each sequence [2–5].

9.4 Cases of the Problem

This problem is investigated in two general cases. In the first case, only the outside ambient temperature is used as the input. In the second case, the date and the sensors' number that show the level of different points from the bottom of the Aub-anbar are used in addition to the ambient temperature [2–5]. In this study, a *MLP* neural network with two layers was used to obtain the temperature profile in the Aub-anbar. The back-propagation algorithm (*BP*) with different training functions and different numbers of neurons in the hidden layer are used for this modeling.

9.4.1 Case # 1

In this case, just the ambient temperature is used as the input of the network and temperature at 36 points at various heights of Aub-anbar are calculated. Since 36 outputs are estimated just from 1 input, the error rate in the results is up to 3°C. The performance of the network can be investigated by considering the regression line's parameters (*R*, *M*, and *B*). Parameter *R* is the correlation coefficient between network outputs and the target. *R* is equal to one in an ideal network that shows a complete correlation between outputs and the targets. *M* and *B* are slope and y-intercept from the best linear regression related to the outputs of the network. In an ideal case, outputs of the network are quite equal to the targets, the slope is equal to 1, and y-intercept is equal to 0.

   Among various training functions, the descent gradient training function and descent gradient training function with momentum have the most error. It is obvious in the scatter plots for modeling values with respect to observational (experimental) values for descent gradient training function and descent gradient training function with momentum (Figures 9.3 and 9.4). The deviation of regression lines from the standard line is relatively high and does not present an acceptable value as a model [2–5]. In these plots, the outputs of the network with respect to the targets are shown in hallow circles. The best linear correlation is shown in a dotted line. The output correlation with respect to the target is shown in a line.

   Descent gradient training function with variable learning rate gives better outputs than two previous training functions (Figure 9.5).

   Descent gradient training function with variable learning rate and momentum presents the best outputs in comparison with three previous functions (Figure 9.6). Therefore, the best training function for this case is the descent gradient training function with variable learning rate and momentum.

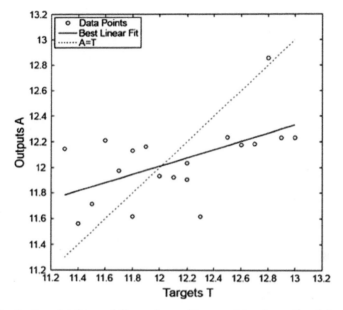

Figure 9.3 Scatter plot for modeling values with respect to observational (experimental) values, descent gradient training function and 30 neurons in hidden layer [2–5].

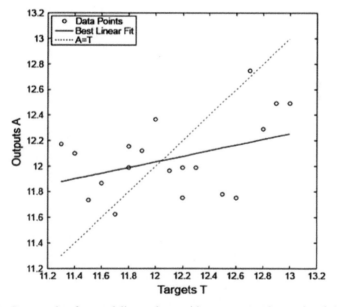

Figure 9.4 Scatter plot for modeling values with respect to observational (experimental) values, descent gradient training function with momentum and 30 neurons in hidden layer [2–5].

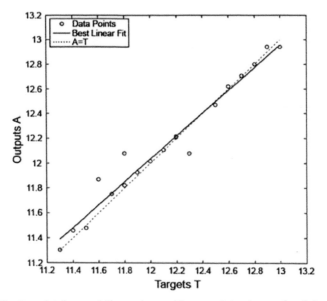

Figure 9.5   Scatter plot for modeling values with respect to observational (experimental) values, descent gradient training function with variable learning rate and 30 neurons in hidden layer [2–5].

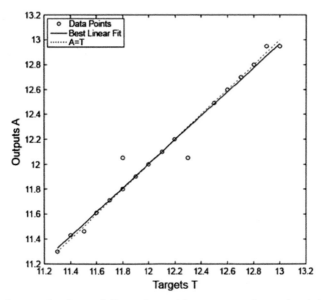

Figure 9.6   Scatter plot for modeling values with respect to observational (experimental) values, descent gradient training function with variable learning rate and momentum and 30 neurons in hidden layer [2–5].

Table 9.1 Various architectures of the network which use the gradient descent method with variable learning rate and momentum as a training function [5]

The Number of Neurons in the Hidden Layer	10	20	30	35	40	50	60
Perf	1601.7	1354.6	556.64	553.8217	553.8217	553.8077	553.8106
M	0.8414	0.8644	0.9665	0.9689	0.9691	0.9693	0.9695
B	1.9004	1.6567	0.9846	0.3757	0.373	0.3709	0.3677
R	0.9255	0.9324	0.9843	0.9846	0.9846	0.9846	0.9846

Pref, performance.

Table 9.1 shows various architectures of the network which use gradient descent method with variable learning rate and momentum as the training function. At the first row of the table, various numbers of hidden layer's neurons are shown. At the second row, performance functions of the network which is sum of squared error between experimental data and the model are shown. As it is clear from the second row, the performance function (sum of squared error) did not decrease for more than 35 neurons in the hidden layer. By increasing hidden layer's neurons more than 35, network's degree of freedom increased and over learning occurred (Figure 9.7).

9.4.2 Case # 2

In this case, the inputs of the network are the ambient temperature on a specific day, the level of considered point from the bottom of Aub-anbar, and the considered date that is specified for measuring the ambient temperature. The output of this network is the temperature at the specified level from the bottom of Aub-anbar at the specified date in the input.

This network estimates one parameter from three parameters. Therefore, the error rate of this network is much lower than a similar network in the previous case with the same training function and equal number of neurons in the hidden layer [2–5].

Regarding the input and output parameters, the number of neurons in the input layer is 3 and the output layer is 1. Among various training functions, the gradient-descent method with variable learning rate and momentum presents a more accurate answer in comparison with other training functions. Also, comparing with the first case, a network similar to the first case was considered in terms of training function, architecture, and variables (except the number of neurons in input and output layers). It was observed that the error rate has decreased considerably and the network gave a better answer compared to the first case.

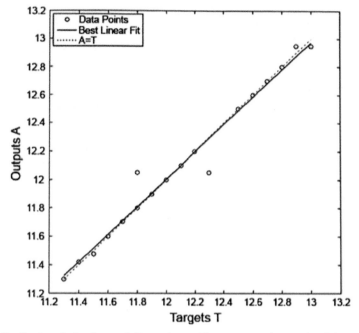

Figure 9.7   Scattered plot for modeling values with respect to observational (experimental) values, descend gradient training function with variable learning rate and momentum with 35 neurons in hidden layer [2–5].

In contrast, a network in the second case that has 5 neurons in the hidden layer and used the gradient descent training function has lower error than a network in the first case that has 35 neurons in the hidden layer and used the gradient descent training function with variable training rate and momentum.

The scatter plots for modeling values with neural networks and various training functions compared to the observational (experimental) values are shown in Figures 9.8–9.11 [2–5].

Table 9.2 presents various architectures of the network that used the training function of gradient-descent method with variable learning rate and momentum. The parameters of this table are similar to the parameters in Table 9.1. As it is clear from the second row, the performance function (sum of squared error) did not decrease for more than 20 neurons in the hidden layer. By increasing hidden layer's neurons more than 20, network's degree of freedom increased and over learning occurred.

Also, when the gradient-descent training function with momentum is used, the number of epochs should be increased. Additionally, there is fluctuation

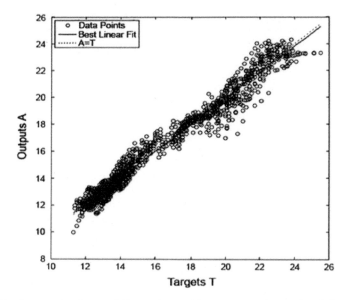

Figure 9.8 Scatter plot for modeling values with respect to observational (experimental) values, gradient descent training function (20 neurons in the hidden layer) [2–5].

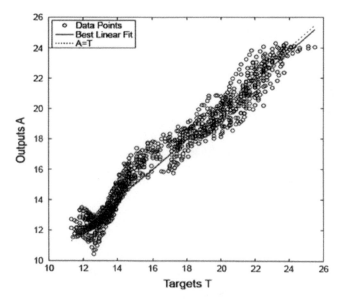

Figure 9.9 Scatter plot for modeling values with respect to observational (experimental) values, gradient-descent training function with momentum (20 neurons in the hidden layer) [2–5].

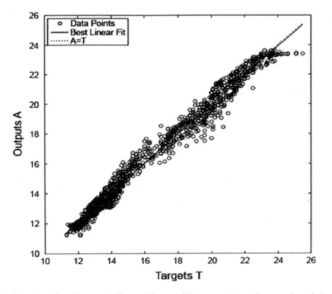

Figure 9.10   Scatter plot for modeling values with respect to observational (experimental) values, gradient-descent training function with variable learning rate (20 neurons in the hidden layer) [2–5].

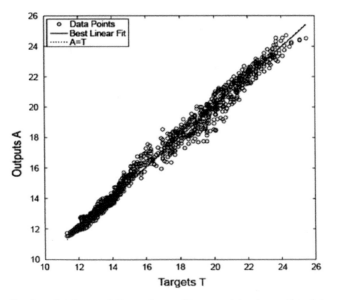

Figure 9.11   Scatter plot for modeling values with respect to observational (experimental) values, gradient-descent training function with variable learning rate and momentum (20 neurons in the hidden layer) [2–5].

Table 9.2   Various architectures of the network which use the gradient-descent method with variable learning rate and momentum as a training function [5]

The Number of Neurons in the Hidden Layer	3	4	5	10	20	35
Perf	1324.4	628.065	301.577	288.866	228.107	262.919
M	0.9248	0.9808	0.9829	0.9839	0.9867	0.9849
B	1.2244	0.3164	0.2753	0.2552	0.2158	0.2448
R	0.9616	0.9906	0.9914	0.9918	0.9935	0.9925

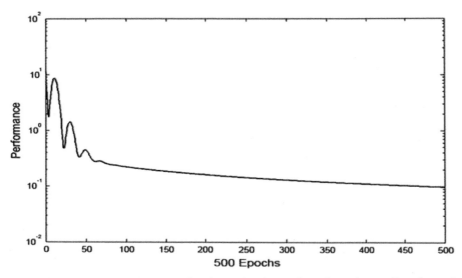

Figure 9.12   Plot of performance function value in number of epochs, gradient-descend training function with momentum (20 neurons in the hidden layer) [5].

in the plot of performance function (sum of squared error) because of the momentum term (Figure 9.12) [2–5]. In this figure, the maximum performance function of the network is equal to 0.097 (the ideal performance number is zero).

9.5 Results of *ANN* Method

In Figures 9.13–9.15, the thermal layer inside Aub-anbar for three specified months in discharge period has been plotted. Thermal layering is shown experimentally and using *MLP* neural network. The networks have 20 neurons in the hidden layer and use the training function of gradient-descent method

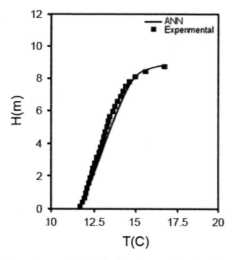

Figure 9.13   Temperature changes in depth for observational and neural network results in the Aub-anbar during June [2–4].

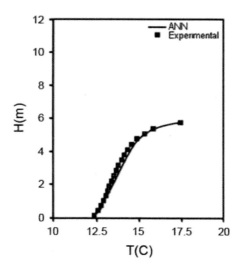

Figure 9.14   Temperature changes in depth for observational and neural network results in the Aub-anbar during August [2–4].

with a variable learning rate and momentum. The back-propagation algorithm method has been used to determine weights and biases of this network. As shown in Figures 9.13–9.15, the temperature profile's difference is low between experimental and neural network [2–5]. As the figures show, the

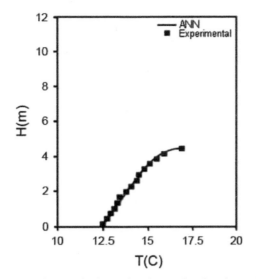

Figure 9.15 Temperature changes in depth for observational and neural network results in the Aub-anbar during September [2–4].

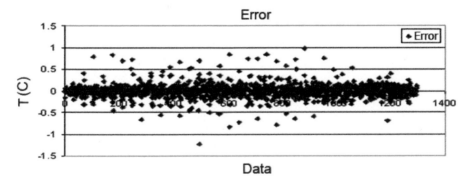

Figure 9.16 Error plot and the difference between observational values and obtained values by *MLP* [2–5].

error rate of this neural network is very low and this method gave a precise answer.

Figure 9.16 shows the error plot that is the difference between obtained temperature from both observation and the neural network on different days and heights of the Aub-anbar. It can be seen that in most of the points, the temperature difference is less than 0.5°C.

At last, by considering the fact that the experimental research is time and money consuming and even impossible in some cases, the *ANN* is a fast, precise, inexpensive, and useful method for the thermal analysis of Aub-anbars in other zones.

References

[1] Menhaj, M. B. (2002). *Fundamentals of Neural Networks*. Amirkabir University of Technology, Tehran.

[2] Madoliat, R., Razavi, M., and Dehghani, A. R. (2009). Modeling of heat transfer in cisterns using artificial neural networks. *J. Thermophys. Heat Transfer* 23.

[3] Razavi, M., Dehghani, A. R., and Khanmohammadi, M. (2009) Simulation of thermal stratification in cisterns using artificial neural networks. *J. Energy Heat Mass Transfer* 31, 201–210.

[4] Ameri Siahoui, H. R., Dehghani, A. R., Razavi, M., and Khani, M. R. (2011). Investigation of thermal stratification in cisterns using analytical and artificial neural networks methods. *J Energy Convers. Manage.* 52.

[5] Razavi, M. (2007). *Modeling of Heat Transfer in Aub-anbars Using Artificial Neural Networks,* MS Thesis, Department of Mechanical Engineering, Islamic Azad University, Science and Research Branch, Tehran, Iran.

[6] Weinbaum, S. (2007). "A Hypothesis for Vulnerable Plaque Ruptures Due to Stress-induced Depending Around Cellular Micro Calcifications in Thin Fibrous Caps," in *the 33rd Annual Northeast Bioengineering Conference, Charles B.* Wang Center, Island, 5–6. 10 and 11 March.

[7] Islamoglu, Y. (2002). *Experimental and Numerical Analysis of Heat Transfer in Corrugated Duct.* Ph.D. Thesis, Sakarya University, Adapazari, Turkey.

[8] Islamoglu, Y., Halici, F., and Parmaksizoglu, C. (2001). "A Computer Aided Analysis of Heat Transfer in Plate Heat Exchanger Channels with Experimental Confirmation," in *Proceedings of Russian National Symposium on Power Engineering*, Kazan, Russia, 256–261.

[9] Diaz, G., Sen, M., Yang, K. T., and McClain, R. L. (2001). Dynamic Prediction and Control of Heat Exchangers Using Artificial Neural Networks. *Int. J. Heat Mass Transfer* 44, 1671–1679.

[10] Parcheco-Vega, A., Diaz, G., Sen, M., Yang, K. T., and McClain, R. L. (2001). Heat rate predictions air–water heat exchangers using correlations and neural networks. *J. Heat Transfer-Trans. ASME 123*, 2, 348–354.

[11] Sablani, S. S., Kacimov, A., Perret, J., Mujumdar, A. S., and Campo, A. (2005). Non-iterative estimation of heat transfer coefficients using artificial neural network models. Int. J. Heat Mass Transfer, 48(3–4), 665–679.

[12] Islamoglu, Y., and Kurt, A. (2004). Heat transfer analysis using ANNs with experimental data for air flowing in corrugated channels. Int. J. Heat Mass Transfer, 47, 1361–1365.

[13] Tsang, T. H., Edgar, T. F., and Hougen, J. O. (1976). *Estimation of heat transfer parameters in a packed bed. Chem. Eng. J.* 11, 57–66.

[14] Kia, M. (2008) *Neural Networks in MATLAB*. Kian, Tehran.

10

Energy and Exergy Analysis in Aub-Anbars

Exergy is defined as the maximum useful work provided by matter or energy, and its evaluation is based on the second law of thermodynamics. According to this law, in every real process, the production of entropy equals with its exergy decay in that process. The main goal of exergy analysis is to identify the location and the amount of non-deliverable various processes of a thermodynamic system. This analysis can help us to investigate the efficiency and performance of the system in question. In exergy analysis, by applying the first and second thermodynamic laws simultaneously, while defining the environment as the reference case, it is tried to answer the questions such as the following:

1. The maximum feasible work obtained from a thermal engine
2. The minimum work required in cooling cycles
3. The feasibility of processes with the least entropy produced.

Since the second thermodynamics law focuses on quality, there is a distinction between the concepts of work and heat [1].

The analysis of the systems based on the second law of thermodynamics or exergy analysis has received attention in the last 3 decades, due to the energy crises in recent years. The fundamentals of exergy analysis theory were developed between 1950 and 1966. The results were the formulas, tables, and graphs to carry out the calculations of exergy analysis in different energy systems.

Assessments of thermal storage systems based on the first law analysis may lead to incomplete results as the essence of any thermal storage unit is to store and deliver proper amount of energy at desired temperature and desired quality. Hence, it is necessary to evaluate the performances of such systems based on both first and second thermodynamics laws.

Exergy analysis has wide applications in the optimization of energy systems. Considering the fact that the book has investigated the heat performance of Aub-anbars, this chapter gives an analysis of energy and exergy in Aub-anbars, as passive systems of heat storage systems. Also, the quality of stored energy in Aub-anbars is investigated. The Aub-anbar under study is the same Aub-anbar mentioned in Chapter 6.

10.1 Energy and Exergy Analysis in the Aub-Anbar Under Study

The temperature distribution during discharge cycle in the reservoir under investigation and the energy and exergy analyses based on the experimental results were carried out to figure out the stored energy and efficiencies of the first and second laws in the reservoir as well. The energy and exergy of the reservoir were calculated every 10 days through the following equations [2, 3]:

$$E = m\,C_{p_w}\,(T_m - T_a) \tag{10.1}$$

$$X = mC_{p_w}\left[(T_m - T_a) - \left(T_a\,\ln\left(\frac{T_e}{T_a}\right)\right)\right] \tag{10.2}$$

where E, X, m, and C_{Pw} are energy, exergy, mass, and specific heat capacity of the stored water, respectively; T_a, ambient temperature; T_m, average temperature of water; and T_e, the equivalent effective temperature of the stored fluid (water).

As long as the temperature variations in the reservoir are considered as a function of height (H), the average temperature of water (T_m), and the equivalent effective stored fluid (T_e); the calculation of exergy is carried out through Equations (10.1) and (10.2) by numerical integral and according to the followings:

$$T_m = \left[\frac{1}{H}\int_0^H T(z)\,dz\right] \tag{10.3}$$

$$T_e = \exp\left[\frac{1}{H}\int_0^H \ln T(z)\,dz\right] \tag{10.4}$$

The energy and the exergy delivered by the outflow or the discharged water are calculated from the following Equations [2, 3]:

$$Q_d = m_e\,C_{p_w}\,(T_{\text{out}} - T_a) \tag{10.5}$$

$$X_d = Q_d - m_e\,C_{p_w}\,T_a\,\ln\left(\frac{T_{\text{out}}}{T_a}\right) \tag{10.6}$$

in which m_e and T_{out} are mass and temperature of outflow during every discharge time step.

It is to be noted that using Equations (10.1) and (10.5) gives negative energy as T_m and T_{out} are below the ambient temperature, and therefore, their absolute values are considered as cold energies. It is also to be noted that T_m is independent of the degree of stratification present in the tank, while T_e depends on it, and in general, $T_e \neq T_m$ except for the condition of fully mixed tank in which T_m is equal to T_e. Therefore, the exergy of the stored and delivered cooled water are dependent on the degree of thermal stratification present inside the tank. The exergies of the water content of the reservoir and discharged water are also calculated for an equivalent fully mixed reservoir by replacing T_e and T_{out} with T_m in Equations (10.2) and (10.6) in order to compare the results with the exergy of the stratified storage. The first and second law efficiencies of the reservoir can also be calculated. The first law efficiency is defined as the ratio of delivered energy (cold energy) to the initial energy content of the tank at the beginning of the discharge period. The second law efficiency is also defined as the ratio of exergy delivered by the reservoir to the initial potential work of the storage tank.

10.2 The Results Obtained from the Analysis of Energy and Exergy

The variation of discharged energy (cold) and exergy with operating or discharge time is presented in Figure 10.1. It is seen that this cold water reservoir is able to store cold energy in winter and deliver it in summer with

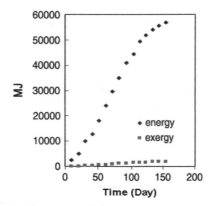

Figure 10.1 Delivered cold energy variation and exergy based on discharge time [2, 3].

the amounts of almost 60 GJ with a temperature of delivered chilled water between 11.5 and 13.1°C. It is also seen that delivered energy although great in quantity is low in quality or work potential.

One of the important features of Aub-anbars is a sure way of delivering drinkable cold water from the beginning of the discharge to the end of that period. As it was mentioned, this is done by filling the reservoir in winter and storing it until the hot seasons. Figure 10.2 presents the obtained cold energy by Aub-anbar during the discharge period which is almost equal to 60 GJ. In other words, 60 GJ cold energy is obtained in winter without any expense and is delivered in summer when the demand is high. The calculations prove that if we want to provide the same amount of energy with refrigerators with 560 w and average coefficient of performance of 1.3, eight of them are needed. The same calculations also prove that we can store 12,820 KWh electrical energy for the same energy stored by Aub-anbar. According to the energy balance sheet reports of 2006, regarding the amount of pollutants produced by power plants (Table 10.1) in Iran, it is observed that the saving of power energy by 12,820 KWh may reduce the amount of pollutants such as CO_2: 7340.77 Kg, SO_2: 12.82 Kg, and NO_x: 11.46 Kg, respectively, in a year [4]. Yazd city, for example, has 93 Aub-anbars [5] which make it possible to save 1192260 *KWh* energy and reduce the pollutants such as CO_2, SO_2, and NO_x up to 68.269 *Ton*, 1192.26, and 1065.78 Kg, respectively, in a year.

Table 10.1 presents the production index of pollutant gases and the green gas in different power plants separately in gr/KWh in 2006 based on the production rate. The maximum production relates to CO_2, SO_2, and NO_x rank the next, respectively.

Green gas is a natural phenomenon which increases the temperature of the surface and the atmosphere of our planet. They include CO_2, NO_2, CFC_s,

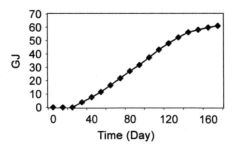

Figure 10.2   The resultant cold energy from Aub-anbar during discharge period [2, 3].

Table 10.1 Index of pollutant and green gases, produced by power plants of Iran in 2007 in gr/KWh [4]

Type Power Plant/Gas	SPM	CH	CO	SO_3	CO_2	SO_2	NO_x	C
Steam	0.126	0.044	0.001	0.020	628.346	1.300	0.973	171.367
Gas	0.145	0.042	0.002	0.019	782.089	1.275	1.252	213.297
Combined Cycle	0.076	0.021	0.001	0.007	487.766	0.469	0.753	133.027
Diesel	0.281	0.091	0.001	0.069	743.178	4.408	1.459	202.685
Hydroelectric	–	–	–	–	6.595	–	–	1.799
Average	**0.108**	**0.034**	**0.001**	**0.015**	**572.603**	**1.000**	**0.894**	**156.165**

O_3, steam, and CH_4. Among those gases, CO_2 comprising 55% has the utmost importance in that it can cause significant changes on the earth. The main source for the production of CO_2 is fossil fuels which have sent about 65% unwanted CO_2 into atmosphere [6, 7].

Figure 10.3 presents the discharged energy and exergy efficiencies of the reservoir under consideration as a function of discharge time. The exergy efficiency of the fully mixed tank with the same average temperature of the stratified tank is also presented for comparison. It is observed that almost 80% of the cold energy stored in the winter is recovered during the summer. The exergy efficiency of the reservoir increases during the discharge time and reaches to a relatively high point at the end of discharge time.

Regarding the balance of exergy, it is seen that a part of available exergy or the primary exergy of the reservoir is retrieved and a part of it remains in

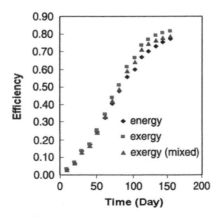

Figure 10.3 The variation efficiencies of the first and second law according to the discharge time of the Aub-anbar under study [2, 3].

the reservoir until the end of discharge period and the rest is wasted. There are three reasons for the wasted exergy:

1. The absorbed energy in the form of heat radiation by the domed roof
2. Heat transfer from or to the soil surrounding the walls and the bottom of the reservoir
3. Convection between the water surface and the air by Baudgeers.

A part of the wasted exergy or decayed exergy relates to internal non-restorability factors such as the transfer of heat from the warm upper layers to the cold bottom layers through convection. The reason for this, as it is shown in Figure 6.5, is the weakening of heat stratification during discharge period.

The reason for high efficiency of exergy at the end of the discharge period is because the design of the cold storing reservoirs is in such a way that they minimize the factors that can cause the waste of energy and exergy.

The efficiency of exergy (mixed) which is shown in Figure 10.3 has been defined as the ratio of exergy or restored work potential of the tank to the initial exergy, where the tank is fully mixed. It is observed that the exergy efficiency of the stratified tank is larger compared to the fully mixed tank. The reason is that the extracted cold energy from the stratified tank is even possible at the lowest level. The high efficiency exergy at the end of discharge period relates to the kind of the design of such reservoirs which minimizes the waste of energy and exergy, as it was mentioned before.

Based on experimental data, the principle of exergy balance in the tank under consideration was studied at the end of discharge period. Table 10.2 shows the elements of exergy: balance of exergy in form of initial exergy of the tank in the beginning of discharge period (X_i), the exergy of the cold water remaining inside the tank at the end of the discharge period (X_f), the exergy recovered from the tank throughout the discharge period (X_r), and, finally, the total wasted and decayed exergy (X_d). Taking all these facts into consideration, it becomes clear that the wasted exergy during the time of cold energy restoring is quite low compared to the initial exergy of the tank and the exergy of the restored cold water which confirm the total high efficiency of the tank's exergy.

Table 10.2   The elements of exergy balance of the tank at the end of discharge period [3]

X_i	X_f	X_r	X_d
2233490	191875	1825370	216245

References

[1] Ghasemi, J. (2001). The performance analysis of the boilers in power plants based on exergy method. MS Thesis, Department of Mechanical Engineering, Tabriz University, Iran.

[2] Dehghan, A. A. (2006). *"Energy and Exergy Analysis of a Seasonal Underground Cold-water Storage,"* in *14th Annual (International) Mechanical Engineering Conference*, Isfahan University of Technology, Isfahan, Iran, May.

[3] Dehghan, A. A., and Dehghani, A. R. (2011). Experimental and theoretical investigation of thermal performance of underground cold-water reservoirs. *Intl. J. Thermal Sci.* 50, 816–824.

[4] *Energy Balance-Sheet in 2006*, Planning Office of Power & Energy, Ministry of Energy, Deputy of Power and Energy Affairs, 2008.

[5] Tavakoli, J. (1991). *Aub-anbars of Yazd*. Cultural Heritage Management, Yazd Province.

[6] Dehghani, M. H. (2005). *Fundamentals of Meteorology and Air Pollution*, 1st edn. Ghashieh, Tehran.

[7] Khani, M. R., Dehghani, A. R., et al. (2007). "Study of Passive Cooling Systems and Chilling Systems Rule in Environmental Pollution Reduction," in *The 1st Conference and Exhibition of Environmental Engineering*, Tehran, Iran, p. 529, Jan.

11

The Pictures of Iran Aub-Anbars

This chapter has provided pictures of the Aub-anbars and ponds in Iran. Due to the climate, Aub-anbars in Iran are both used in hot, arid and hot, humid regions. These structures have been classified according to their importance and antiquity.

11.1 Aub-Anbars of Hot, Arid Regions

City of Yazd	Province of Yazd
City of Ardakan	Province of Yazd
City of Meybod	Province of Yazd
City of Taft	Province of Yazd
City of Abarkooh	Province of Yazd
City of Naein	Province of Isfahan
City of Jandagh	Province of Isfahan
City of Kashan	Province of Isfahan
City of Ardestan	Province of Isfahan
City of Kooh Payeh	Province of Isfahan
City of Harand	Province of Isfahan
City of Isfahan	Province of Isfahan
City of Rafsanjan	Province of Kerman
City of Anar	Province of Kerman
City of Bardaskan	Province of Khorasan Razavi
City of Kashmar	Province of Khorasan Razavi
City of Sabzevar	Province of Khorasan Razavi
City of Ferdows	Province of Khorasan Razavi
City of Ghaen	Province of Khorasan Razavi
City of Birjand	Province of South Khorasan
City of Khosf	Province of South Khorasan
City of Nehbandan	Province of South Khorasan

City of Garmsar	Province of Semnan
City of Ghazvin	Province of Ghazvin
City of Larestan	Province of Fars

11.2 Aub-Anbars of Hot, Humid Regions

Qeshm Island	Province of Hormozgan
Bandar-e Lengeh	Province of Hormozgan
Bandar-e Kong	Province of Hormozgan
City of Dejegan	Province of Hormozgan
Bandar-e Moallem	Province of Hormozgan
Bandar-e Homeiran	Province of Hormozgan
Bandar-e Khamir	Province of Hormozgan
Bandar-e Magham	Province of Hormozgan
City of Kokhard	Province of Hormozgan
City of Parsian	Province of Hormozgan
Kish Island	Province of Hormozgan
Bandar-e Kangan	Province of Bushehr
City of Sari	Province of Mazandaran

Figure 11.1 A view of reservoir and two-sided Baudgeers of Dowlat Abad Aub-anbar in Yazd, Yazd Province.

Figure 11.2 A picture of Asr Abad Aub-anbar with seven Baudgeers in near Yazd, Yazd Province.

Figure 11.3 A view of "Six-Baudgeer Aub-anbar" in Yazd, Yazd Province.

Figure 11.4   A view of entrance, Baudgeers and domed roof of "Six-Baudgeer Aub-anbar" in Yazd, Yazd Province [1].

Figure 11.5   A view of Rostam Giev Aub-anbar in Yazd, Yazd Province [2].

Figure 11.6 A view of Hojjat Abad Aub-anbar with eight-sided Baudgeers in the route of Yazd to Maybod, Yazd Province [2].

Figure 11.7 A view of four Baudgeers and reservoir of Vaght-o Saat Aub-anbar in Yazd, Yazd Province.

Figure 11.8   A view of short six-sided Baudgeers and domed roof of Aub-anbar in Yazd, Yazd Province.

Figure 11.9   A picture of six-sided Baudgeers of Aub-anbar in Yazd, Yazd Province.

Figure 11.10 An Aub-anbar beside the Zoroastrians' dungeon in Yazd, Yazd Province.

Figure 11.11 A picture of Baudgeers and domed roof of "Five-Baudgeer Aub-anbar" of the Amir Chakhmagh Complex in Yazd, Yazd Province.

Figure 11.12   Two-sided Baudgeers of the Shohada-ye-Fahraj's Aub-anbar (part of the structure is underground) in Yazd, Yazd Province.

Figure 11.13   Naserieh Aub-anbar beside Khan garden in Yazd, Yazd Province.

Figure 11.14 An Aub-anbar enjoying seven Baudgeers as well as two reservoirs in Hossein Abad village near Yazd, Yazd Province.

Figure 11.15 Pir-e herisht Aub-anbar in Ardakan City, Yazd Province [2].

Figure 11.16   An Aub-anbar with one-sided Baudgeers in Ardakan City, Yazd Province.

Figure 11.17   Akhound Aub-anbar in Ardakan City, Yazd Province.

Figure 11.18   Shahri-ha Aub-anbar in Ardakan City, Yazd Province.

Figure 11.19   An Aub-anbar in Aghda, Ardakan City, Yazd Province.

Figure 11.20   A view of Kolar Aub-anbar with one-sided Baudgeers in Maybod City in Yazd Province.

Figure 11.21   A view of one-sided Baudgeers and domed roof of Caravanserai Aub-anbar in Maybod City, Yazd Province.

Figure 11.22 An Aub-anbar of brick-made in Taft City, Yazd Province.

Figure 11.23 En route Aub-anbar beside Zein Abad Village near the route of Yazd to Taft, Yazd Province.

Figure 11.24   A view of Aub-anbar near Abarkooh City, Yazd Province.

Figure 11.25   A view of three Baudgeers and entrance of Mosalla Aub-anbar in Naein City which is outside the garden, Isfahan Province.

Figure 11.26   A view of Massoum Khani Aub-anbar in Naein City, Isfahan Province.

Figure 11.27   An Aub-anbar in Naein City with two reservoirs and several Buadgeers being restored, Isfahan Province.

Figure 11.28   A view of Aub-anbar in Mohammadieh, Naein City, Isfahan Province.

Figure 11.29   A view of entrance, domed roof, and three Baudgeers of Aub-anbar in Naein City, Isfahan Province.

Figure 11.30 A picture of Aub-anbar in Mohammadieh of Naein City, Isfahan Province.

Figure 11.31 An Aub-anbar in Mesr Village, Jandagh, Naein City, Isfahan Province.

Figure 11.32   Khan Aub-anbar in Kashan City, Isfahan Province.

Figure 11.33   Panjeh Shah Aub-anbar in Kashan City, Isfahan Province.

Figure 11.34 An Aub-anbar in around Kashan City, Isfahan Province.

Figure 11.35 Haj Hassan Aub-anbar with two reservoirs and four four-sided Buadgeers in Ardestan City, Isfahan Province.

Figure 11.36 An Aub-anbar beside Shah Abbasi Caravanserai in Kooh Payeh City, Isfahan Province.

Figure 11.37 An Aub-anbar in Toushtak, Kooh Payeh City, Isfahan Province.

Figure 11.38 An Aub-anbar in Harand City, Isfahan Province (one of its Baudgeers has partially damaged).

Figure 11.39 A view of Kazerooni Aub-anbar in Takht-e Foulad, Isfahan Province.

Figure 11.40   An Aub-anbar in Brahman a suburb of Rafsanjan City, Kerman Province.

Figure 11.41   Aub-anbar of Hossein Abad, in Anar City, Kerman Province.

Figure 11.42 Seyed Bagher Aub-anbar in Bardaskan City, Khorasan Razavi Province.

Figure 11.43 A view of Aub-anbar of Lotf Abad Fadafan in Kashmar City, Khorasan Razavi Province.

Figure 11.44 A view of two Baudgeers and entrance of Mazinan in Sabzevar City, Khorasan Razavi Province.

Figure 11.45 A view of several Aub-anbar in Namen of Sabzevar City, Khorasan Razavi Province.

Figure 11.46   Joogi Aub-anbar in Ferdows City, Khorasan Razavi Province.

Figure 11.47   Karshak Aub-anbar in Ghaen City, Khorasan Razavi Province.

Figure 11.48   A view of Aub-anbar of Amir Abad in Birjand, South Khorasan Province.

Figure 11.49   An Aub-anbar in Khosf City, South Khorasan Province.

Figure 11.50 Mighan Aub-anbar in Nahbandan City, South Khorasan Province.

Figure 11.51 An Aub-anbar in Ali Abad, Garmsar City, Semnan Province.

Figure 11.52   A view of brick-cover Aub-anbar in Garmsar City, Semnan Province.

Figure 11.53   A view of the single Buadgeer of Haj Kazem Aub-anbar in Ghazvin, Ghazvin Province.

Figure 11.54   A view of entrance of Sardar Bozorg Aub-anbar in Ghazvin, Ghazvin Province.

Figure 11.55   A view of reservoir of Sardar Bozorg Aub-anbar in Ghazvin, Ghazvin Province.

Figure 11.56   A view of several Aub-anbars in Larestan, Fars Province.

Figure 11.57   An Aub-anbar in Larestan, Fars Province.

Figure 11.58 Bibi Aub-anbar in Gheshm Island, Hormozgan Province.

Figure 11.59 An Aub-anbar renovated in Gheshm Island, Hormozgan Province.

Figure 11.60   A view of two Aub-anbars in Gheshm Island, Hormozgan Province [3].

Figure 11.61   A view of Aub-anbar in Bandar-e Lengeh, Hormozgan Province.

Figure 11.62 A view of Chahar Berke (or star-reservoir) in Bandar-e Kong, Hormozgan Province.

Figure 11.63 A view of long and circular Aub-anbars in Dejegan City, Hormozgan Province.

Figure 11.64   An internal view of a long Aub-anbar in Dejegan City, Hormozgan Province.

Figure 11.65   A view of Chahar Berke (or star reservoir) in Bandar-e Moallem, Hormozgan Province.

Figure 11.66 A view of Aub-anbar with conic roof in Bandar-e Homeiran, Hormozgan Province.

Figure 11.67 A view of Aub-anabr with a white conic roof in Bandar-e Homeiran, Hormozgan Province.

Figure 11.68   A view of Aub-anbar in Bandar-e Khamir, Hormozgan Province.

Figure 11.69   An Aub-anbar in Bandar-e Magham, Hormozgan Province [3].

Figure 11.70 A view of small Aub-anbar in Kokhard City, Hormozgan Province.

Figure 11.71 A view of Aub-anbar in Parsian City, Hormozgan Province.

Figure 11.72   An Aub-anbar in Kish Island, Hormozgan Province.

Figure 11.73   A view of Aub-anbar in Bandar-e Kangan, Bushehr Province.

Figure 11.74   An Aub-anbar with clay cover in Bandar-e Kangan, Bushehr Province.

Figure 11.75   A view of Aub-anbar in Sari City, Mazandaran Province.

References

[1] J. Ghasbanpour, *Modern Life, Old Body; an Anthology of Precious Historical Architects*, Tehran: Ministry of Housing and City Development with the Cooperation of Iranian Cultural Heritage Organization, 2nd Edition, 1995.

[2] H. Amir Yeganeh, *Yazd, the Gem of the Desert*, Tehran: Honar Sarai-e Goya, 2005.

[3] N. Kasraeian, *Iran, Our Land*, Tehran: Nasrollah Kasraeian, 1993.

Index